EXPLORATIONS OF MATHEMATICAL MODELS IN BIOLOGY WITH MATLAB®

EXPLORATIONS OF MATHEMATICAL MODELS IN BIOLOGY WITH MATLAB®

MAZEN SHAHIN
Department of Mathematical Sciences
Delaware State University
Dover, DE

WILEY

Published by John Wiley & Sons, Inc., Hoboken, New Jersey
Published simultaneously in Canada

Library of Congress Cataloging-in-Publication Data

Shahin, Mazen, 1947–
Explorations of mathematical models in biology with MATLAB / Mazen Shahin, Department of Mathematical Sciences, Delaware State University, Dover, DE.
pages cm
Includes bibliographical references and index.
ISBN 978-1-118-03212-1 (cloth)
1. Biology–Mathematical models. 2. MATLAB. I. Title.
QH323.5.S465 2013
570.1′5195–dc23

2013026539

10 9 8 7 6 5 4 3 2 1

Dedicated to Nina, Ramy, and Randa

CONTENTS

PREFACE

MAIN GOALS

The main goal of *Explorations of Mathematical Models in Biology with MATLAB®* is to offer students a textbook to help them explore and discover mathematical concepts and use those concepts in building and analyzing mathematical models of life science disciplines such as biology, ecology, and environmental sciences. The main mathematical tools used in this text are difference equations and matrices. The use of the mathematics software MATLAB is an integral part of exploring and analyzing the models. It is important to stress that this text is neither a text on difference equations nor a text on how to use MATLAB.

THE NEED FOR A TEXTBOOK IN ELEMENTARY MATHEMATICAL MODELING AND MATRIX ALGEBRA

In 2005, the Mathematical Association of America (MAA) published *Math & Bio 2010: Linking Undergraduate Disciplines. Math & Bio 2010* envisions a new educational paradigm in which the disciplines of mathematics and biology, currently quite separate, will be productively linked in the undergraduate science programs of the twenty-first century. As a science, biology depends increasingly on data, algorithms, and models; in virtually every respect, it is becoming more quantitative, more computational, and more mathematical.

One of the recommendations of the Curriculum Renewal Across the First Two Years (CRAFTY) committee of the MAA published in 2004 was the inclusion of mathematical modeling, discrete mathematics, and matrix algebra in the mathematics curriculum for biology majors.

Following recommendations of these professional organizations, we believe that there is a need for a course in elementary mathematical modeling and matrix algebra for biology and life science majors. Traditionally, mathematical models use differential equations. A differential equations course is usually offered after calculus I and calculus II. Consequently, there is a great need to offer a course in mathematical modeling that uses difference equations rather than differential equations. Difference equations and matrix algebra are simple yet powerful tools in modeling discrete time dynamical systems. These tools are accessible to students with high school algebra II or college algebra and do not require calculus and differential equations.

Because one of the main objectives in writing this book is to provide students with a self-contained textbook, we included the background materials necessary to understand the main topics in the text. For example, we present necessary materials from linear algebra.

APPROACH

In this text, we introduce the modeling of real-life situations with difference equations and matrices using MATLAB. We emphasize the use of graphical and numerical techniques, rather than theoretical techniques, to investigate and analyze the behavior of solutions of the mathematical models. We investigate interesting linear and nonlinear models from diverse life science disciplines, such as biology, ecology, and environmental sciences.

We use a discovery pedagogical approach. To introduce a concept, first we investigate a model numerically and/or graphically and recognize a pattern or certain properties that characterize that concept. Then we give a definition of the concept with examples and applications. For example, to introduce the eigenvalues and corresponding eigenvectors of a square matrix we investigate an age-structure population model with different initial population vectors that lead to visualization of the eigenvalues and corresponding eigenvectors. Then the definition of the eigenvalues and eigenvectors are introduced and some properties are discussed.

WHY MODELING WITH DIFFERENCE EQUATIONS AND MATRICES?

Difference equations represent a very sophisticated and powerful mathematical tool for modeling a wide range of real-life discrete time situations in diverse areas, including the life sciences. And matrices provide an excellent tool in modeling linear problems. Moreover, these powerful tools do not require a sophisticated mathematics background and are accessible to anyone who has successfully completed high school or college algebra.

WHY DO WE USE MATLAB?

All the models presented in the text require the use of computers. For example, to investigate and analyze a model it is often necessary to iterate the difference equation(s) or the matrix difference equation(s) that represent the model and graph it. Sometimes it is necessary to find the eigenvalues and the corresponding eigenvectors to investigate the long-term behavior of a dynamical system. In other instances, to investigate the sensitivity of a dynamical system to certain parameters, it is necessary to change the parameters of the dynamical system and find the corresponding numerical solutions. All these computational activities require software that is easy to learn and use. In this text we use MATLAB, which is a very user-friendly and powerful mathematics program with excellent graphing capabilities. The use of MATLAB frees students from tedious calculations. This allows them to focus on translating a problem into mathematical notation, finding a solution, interpreting the numerical and the graphical information provided, making conjectures, and writing about their findings and observations. With the use of MATLAB the students focus on building and analyzing the models.

ORGANIZATION

The materials in the text follow a logical order. Chapter 1 is an introduction to the mathematical modeling process, basic difference equations terminology, and getting started with MATLAB. Chapter 2 introduces modeling with linear and nonlinear first-order difference equations. Sections 2.1 and 2.2 focus on models with linear difference equations, such as population dynamics of a single species, the concentration of a drug in the bloodstream, radioactive decay and carbon dating, and forensic applications of Newton's law of cooling. Section 2.3 introduces modeling with nonlinear difference equations; logistic growth models with and without harvesting are investigated. Section 2.4 is an intuitive introduction to chaos.

Chapter 3 introduces basic concepts of matrix algebra. Section 3.1 introduces systems of linear equations having unique solutions and uses models on nutrition. In Section 3.2 we discuss the Gauss-Jordan elimination methods by modeling allocation of resources, balancing chemical equations, and determining the time of death of a murder victim. In Section 3.3 we introduce the standard matrix notation and the basic matrix operations, such as addition, subtraction, scalar multiplication, multiplication, and the inverse of a square matrix. Section 3.4 is a simple introduction to the main concepts of determinants and their role in finding the inverse of a matrix and determining whether a system of linear equation has a unique solution, many solutions, or no solution. In Section 3.5 we intuitively introduce the concept of eigenvalues and eigenvectors and investigate methods for determing the eigenvalues of a square matrix and the associated eigenvectors. In Section 3.6 we investigate the use of eigenvalues for determining the long-term behavior of a system of linear equations.

Chapter 4 concentrates on modeling with systems of linear difference equations. In Section 4.1 we discuss a special class of finite stochastic processes and modeling with Markov chains. Section 4.2 provides investigations of Leslie's age-structured population models. In Section 4.3 we investigate modeling real-life situations with second-order linear difference equations; examples include seal population dynamics and a plant population model. The eigenvalues and eigenvectors are efficiently used as a tool for studying the long-term behavior of the models introduced in the chapter.

Chapter 5 is an introduction to modeling with nonlinear systems of difference equations. Section 5.1 is devoted to investigating the dynamics of interacting species, such as predator–prey and competing species. In Section 5.2 we investigate some models of the spread of contagious diseases, such as the SIR and SIS models. Section 5.3 considers some models represented by second-order nonlinear difference equations, such as delayed logistic models.

THE INTENDED AUDIENCES

The intended audiences are life sciences, mathematics, and mathematics education majors. The life sciences include biology, ecology, environmental science, and forensic science. The text can serve as a mathematics course in the liberal arts core or as a general education requirement mathematics course. It can also serve as an honors mathematics course for all majors.

SUPPORT WEBSITE

Supplementary material for this book can be found by entering ISBN 9781118032121 at booksupport.wiley.com. This website contains solutions to selected problems, worksheets, and MATLAB code for many programs used in the text, including so-called M-files.

ACKNOWLEDGMENTS

I am indebted to Eric Frankl for proofreading the entire manuscript. Eric caught and corrected several typos and mistakes. I am grateful to Tabitha Myers, who typed several parts of this book. I am thankful for Saber Elaydi who advised and encouraged me to write this book. I am indebted to Nancy Baxter Hastings who consulted me on early prepublished versions of this text and greatly supported me. Thanks to Leah Hontz for helping in forming some exercises.

I would like to express my gratitude for the people at John Wiley & Sons Inc. who worked with me on the development of this text. Special thanks to the Mathematics and Statistics Senior Editor, Susanne Steitz-Filler, and the Senior Editorial Assistant of Mathematics and Statistics, Sari Friedman.

CHAPTER 1

OVERVIEW OF DISCRETE DYNAMICAL MODELING AND MATLAB®

1.1. INTRODUCTION TO MODELING AND DIFFERENCE EQUATIONS

In this section we introduce dynamical systems, discuss discrete dynamical systems vs. continuous dynamical systems and informally define a mathematical model.

1.1.1. Model 1.1: Population Dynamics, A Discrete Dynamical System

Consider the population of a city with a constant growth rate per year. The population is counted at the end of each year. For simplicity, assume that there is no immigration to or emigration from the city.

i. Model the population dynamic and predict the long-term behavior of the system.
ii. In 2010, the city's population was 100,000. The natural annual growth rate of the population is 1% per year. Predict the city's population in 2020. Estimate the population over the next 30 years and graph it. What is the long-term behavior of the population?

Explorations of Mathematical Models in Biology with MATLAB®, First Edition. Mazen Shahin.

Discussion

i. We will measure the population at discrete time intervals in one-year units. Let

p_n = population size at the end of the time period (year), n.
p_0 = the initial population size, 0.
r = the constant growth rate per period (year).

The relationship between the current population, p_n, and the next population, p_{n+1}, is

$$p_{n+1} = p_n + rp_n.$$
$$p_{n+1} = (1+r)p_n. \tag{1.1}$$

Therefore, the population dynamics can be modeled by equation 1.1.

Equation 1.1 is a **difference equation** (or recurrence equation). The system 1.1 and the initial value, p_0, represent the population dynamics. Because the population changes over time, this system is a **dynamical system**. Because this dynamical system changes over discrete time intervals, the system is called a **discrete dynamical system**. We say that the population dynamics is **modeled** by the discrete dynamical system (or the difference equation 1.1).

To find p_k, use p_0 in equation 1.1 to find p_1, then use p_1 to find p_2 and so on until p_k. This process is called **iteration** of the difference equation 1.1, and the sequence 1.2,

$$p_0, p_1, p_2, \ldots, p_k, \tag{1.2}$$

for any value of k (positive integer) is called a **solution** or **numerical solution** of the given difference equation 1.1.

From equation 1.1, if the current value of p_n is known, the next value, p_{n+1}, can be calculated. For example, if we have p_5 we can calculate p_6. However, if we have p_0, equation 1.1 does not allow us to calculate, for example, p_6 in one step. Therefore, we are in need of a closed form to calculate p_n in one step if we know the values of p_0 and n. It can be easily proven that

$$p_n = (1+r)^n p_0. \tag{1.3}$$

Equation 1.3 is called the **analytical solution** of the difference equation 1.1.

Equation 1.3 is an exponential function and will grow or decay exponentially, depending on the value of r. If $r>0$, then $(1+r)>1$, and therefore population size, p_n, grows unbounded when n is very large. If $r<0$, then $(1+r)<1$, and consequently, the population size approaches zero when n is very large.

ii. Let's apply the model just discussed to the given information, where $r = 0.01$ and $p_0 = 100{,}000$. The city's population is modeled by the system

$$p_{n+1} = 1.01p_n, \quad p_0 = 100{,}000. \tag{1.4}$$

To find the population in 2020 (10 years from 2010), we use equation 1.3 with $n = 10$ and $p_0 = 100{,}000$. We are looking for p_{10}. We have

$$p_{10} = (1.01)^{10}100{,}000 = 110{,}462.$$

One way to find the values of $p_1, p_2, \ldots, p_{30}$ is to iterate equation 1.4. Then graph the ordered pairs (n, p_n). To illustrate how the iteration works, let's see how to calculate, for example, p_3. Set $n = 0$ in equation 1.4 to get p_1,

$$p_1 = 1.01p_0 = 1.01(100{,}000) = 101{,}000.$$

Then set $n = 1$ in equation 1.4 to get p_2,

$$p_2 = 1.01p_1 = 1.01(101{,}000) = 102{,}010.$$

Finally, setting $n = 2$ in equation 1.4 gives p_3,

$$p_3 = 1.01p_2 = 1.01(102{,}010) = 103{,}030.$$

In particular, to find P_{10}, the answer to the original question, we can use the difference equation 1.4 and the initial condition to find the sequence $p_0, p_1, p_2, \ldots, p_{10}$ which is 100,000; 101,000; 102,010; 103,030; 104,060; 105,101; 106,152; 107,213; 108,285; 109,368; 110,462. Thus $p_{10} = 110{,}462$.

Usually we use MATLAB to iterate a difference equation. We will introduce MATLAB in Section 1.3. The graph of p_n, $n = 0, 1, 2, \ldots, 30$ vs. n is shown in Figure 1.1.

The analytical solution and numerical solution of equation 1.4 show that the city population slowly grows unbounded as n becomes very large.

1.1.2. Model 1.2: Population Dynamics, A Continuous Dynamical System

Consider the following situation. There are some bacteria in a tube with a nutritive solution. As time progresses, the bacteria reproduce by splitting and dying. Assuming that there is enough food and space for the bacteria, model the dynamics of the bacteria. Investigate the long-term behavior of the model.

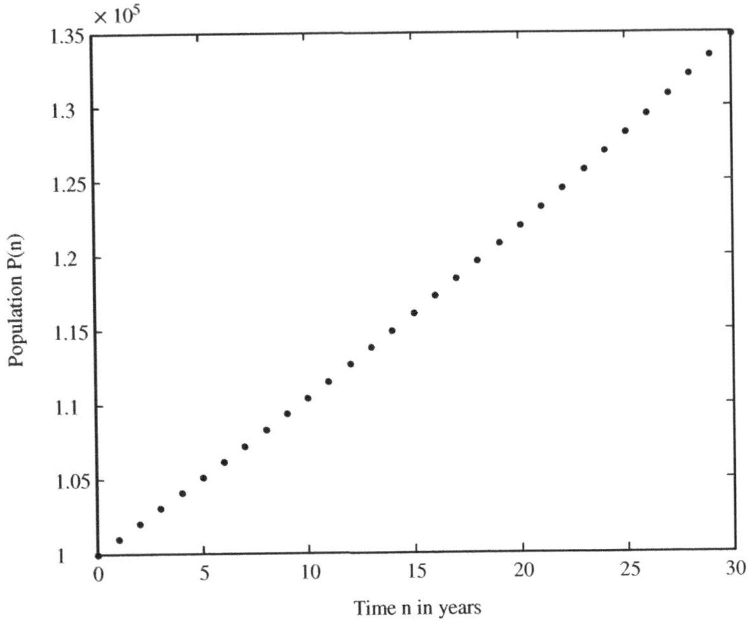

FIGURE 1.1. Graph of a city's population after n years, p_n, vs. time, n, in years. The population is modeled by the difference equation $p_{n+1}=1.01p_n$, with the initial population $p_0=100{,}000$ and $n=0, 1, \ldots, 30$.

Discussion

Let $p(t)$ be the bacteria's population size (number of bacteria) at time t and $p(0)=p_0$. Assume that the growth rate $r=b-d$, where b is the birth rate and d is death rate. The assumption that there is enough food and space means there is no restriction on the increasing number of bacteria. Therefore, the rate of change of bacteria's population size $\frac{dp}{dt}$ is proportional to the bacteria's population p. Consequently, the dynamic of the bacteria is modeled by the dynamical systems 1.5 and 1.6:

$$\frac{dp}{dt}=(b-d)p=rp. \tag{1.5}$$

$$p(0)=p_0. \tag{1.6}$$

Equation 1.5 is a **first-order ordinary differential equation** and equation 1.6 is called the initial value (condition). From the basics of differential calculus or differential equations, the solution of systems 1.5 and 1.6 is

$$p(t)=e^{rt}p_0. \tag{1.7}$$

Knowing the values of r and p_0, the population, $p(t)$, can be evaluated at any time t. For $r>0$, equation 1.7 implies that the population size, $p(t)$, increases and grows unbounded as $t \to \infty$, while $r<0$ implies that the population size decreases and approaches zero as $t \to \infty$.

Because the change in the system 1.5 is continuous, this system is called a **continuous dynamical system**. Usually continuous dynamical systems are represented by one or more ordinary or partial differential equations. Note that the models discussed in this text are restricted to discrete dynamical systems. In other words, the text discusses only those models represented by difference equations.

1.1.3. Why Modeling with Difference Equations is Adopted

There are compelling reasons to restrict models in this text to difference equations, including

1. Modeling with difference equations is a very powerful, yet simple tool for modeling dynamical systems in biology, ecology, the environment, and chemistry.
2. Modeling with difference equations requires knowledge of algebra but does not require knowledge of differential calculus; modeling with differential equations requires a course in differential equations. This text targets freshman and sophomores majoring in the life sciences and mathematics, many of whom have not had a differential equations course.

1.1.4. What is a Mathematical Model?

The following is a possible informal definition of a mathematical model:

> A mathematical model is a translation of a real-world problem into mathematical notation by forming a mathematics problem corresponding to the real-world problem. Then mathematics tools, ideas, concepts, and techniques are used to solve the mathematics problem. The obtained solution is translated back into the real-world problem context.

1.1.5. Basic Terminology of Difference Equations

The **order** of a difference equation is equal to the difference between the largest and the smallest indices in the difference equation. For example, in the difference equations 1.1 and 1.4, the difference between the largest subscript, $n+1$, and the smallest subscript, n, is 1. Therefore each of equations 1.1 and 1.4 is a **first-order difference equation**. The difference equation 1.8 is a **second-order difference equation**,

$$x_{n+2} = x_{n+1} + x_n. \tag{1.8}$$

Examples of first-order difference equations are

$$y_n = 0.8y_{n-1} + 20. \tag{1.9}$$

$$p_n = 1.02p_{n-1}. \tag{1.10}$$

$$x_{n+1} = 1.2x_n - 0.15n. \tag{1.11}$$

$$z_n = n^2 z_{n-1} + 2^n. \tag{1.12}$$

$$B_{n+2} = \frac{4}{n}B_{n+1} - n^{1/2}. \tag{1.13}$$

$$y_{n+1} = 3y_n^2 + 3n. \tag{1.14}$$

$$x_n = \sqrt{x_{n-1}} - 23. \tag{1.15}$$

$$y_{n+1} = y_{n+1}y_n + 6. \tag{1.16}$$

A difference equation is called **linear** if its terms are not raised to a power other than 1 and if the terms are not multiplied together. Otherwise the difference equation is called **nonlinear**. For example, the difference equations 1.1, and 1.8–1.13 are linear. The difference equation 1.14 is nonlinear because y_n is raised to the power of 2. Equation 1.15 is nonlinear because x_{n-1} is raised to the of power $\frac{1}{2}$. In equation 1.16, two terms, y_{n+1} and y_n, are multiplied together; therefore it is nonlinear.

A **first-order linear difference equation** can be represented by the equation

$$y_{n+1} = a_n y_n + b_n, \tag{1.17}$$

where a_n and b_n are two known sequences, and a_n is a coefficient. Note that each of the sequences a_n and b_n can be a constant sequence or can depend on n. When the sequence a_n is a constant, such as a, the difference equation

$$y_{n+1} = ay_n + b_n \tag{1.18}$$

is called a **first-order linear difference equation with constant coefficients**. When $b_n = 0$, the difference equation is called **homogenous** and equation 1.17 becomes

$$y_{n+1} = a_n y_n. \tag{1.19}$$

Equation 1.19 is called a **first-order linear homogenous difference equation**. If $a_n = a$, where a is a constant, equation 1.19 becomes

$$y_{n+1} = ay_n \tag{1.20}$$

and is called a **first-order linear homogenous difference equation with constant coefficients**. For example, the difference equations 1.1, 1.4 and 1.10 are first-order linear homogenous difference equations with constant coefficients, and equations 1.9 and 1.11 are first-order linear difference equations with constant coefficients.

1.1.6. Exercises 1.1

1. Suppose that the initial population of a species is 10,000 and the growth rate is 5% per year.

A. Model this situation by a difference equation.
B. Find the first five populations.
C. Find the population size after 20 years.

2. Assume that the population of a country is 100 million and the natural growth rate is 2% per year. Assume that 20,000 immigrants are allowed into the country every year.

A. Model this situation by a difference equation.
B. Find the population for the next 4 years.
C. Find the population after 10 years.
D. Can you predict the long-term behavior of the system?

3. Assume that the kidneys remove 25% of an anesthetic substance from the body every hour. A patient is injected with 600 mg of the substance before a minor surgery on her tooth.

A. Model this situation by a difference equation.
B. Determine the amount of the substance in the patient's body after the first 4 hr.
C. What is the long-term behavior of the system?

4. In 2000, a lake contained 800 lb of contamination. In the same year, a new plant started to dump 120 lb of the contaminant into the lake every year. Assume that 15% of the contaminant in the lake is naturally removed from the lake every year. Assume that this trend continues.

A. Model the amount of contaminant in the lake at any year after 2000.
B. Determine the amount of the contaminant in the lake in years 2001–2006.
C. Can you determine the long-term behavior of the system?

5. Identify the following difference equations as linear or nonlinear.

A. $x_{n+1} = 3x_n + 2$
B. $y_n = 2y_{n-1} - 4n$
C. $y_{n+1} = 4\sqrt{y_n} + 5$
D. $x_{n+1} = x_n x_{n-1} + 6$
E. $z_{n+2} = 2z_{n+1} - 3$
F. $x_{n+1} = 3x_n^2 + 4$

6. Determine the order of the following difference equations.

A. $x_{n+1} = 2x_n + 6n$
B. $x_{n+1} = 4x_n - 2x_{n-1}$

C. $y_{n+2} = 4\sqrt{y_n} + 5$

D. $z_{n+2} = 2nz_{n+1} - 4z_n$

E. $z_{n+2} = 2z_{n-1} - 3$

F. $y_n = y_{n-2}^2 + 4$

7. Determine whether the following difference equations are homogenous or nonhomogenous

A. $x_{n+1} = 3x_n + 6n$

B. $y_{n+2} = 4y_{n+1} - 2y_n$

C. $x_{n+1} = 2x_n + n^2 + 5n + 2$

D. $y_{k+3} = 5y_{k+1} + 6$

E. $z_n = n^2 z_{n-1} - nz_{n-2}$

F. $z_{n+2} = n^3 z_{n+1} + 4z_n + 2n$

8. Identify the order, linearity, and homogeneity of each of the following difference equations.

A. $x_{n+3} = 2x_{n+1} + 3x_n - n^2$

B. $y_{n+2} = 4x_{n+1}x_n - 4$

C. $z_{n+1} = 3z_n + 4n$

D. $P_{n+2} = 2nP_n$

E. $P_{n+1} = P_n + aP_n - bP_n^2$, where a and b are constants.

F. $x_{n+1} = rx_n(1 - x_n)$, where r is a constant

1.2. THE MODELING PROCESS

It is useful to view mathematical modeling as a process, as illustrated in Figure 1.2 The modeling process is represented by a loop, where the starting point is Step 1, located in the upper-left-hand corner of the figure.

1.2.1. Step 1: Formulate the Real-World Problem

The first step is to get information pertaining to the system under consideration and identify the question(s) to be answered. The question should be neither too general nor too narrow. If the formulated question is too general, it is difficult to manage the problem; if the question is too narrow, the problem might become trivial. Because the question will be translated into mathematical notation, it should be stated in precise mathematical terms.

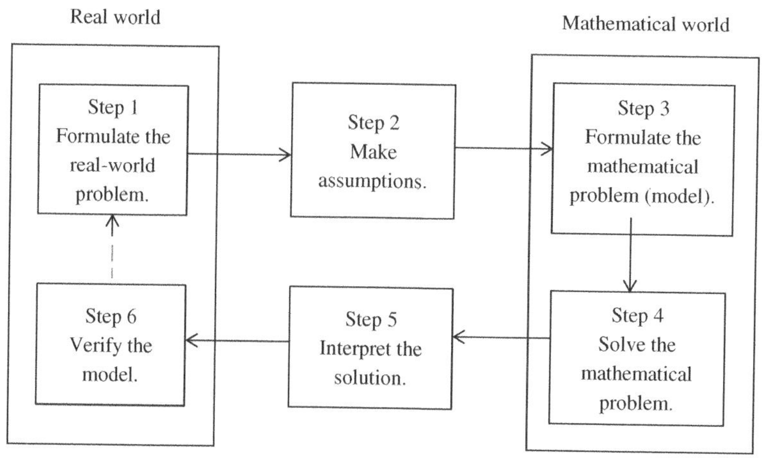

FIGURE 1.2. The mathematical modeling process.

As an example we consider two species in a forest, rabbits and foxes, where foxes eat rabbits and there is enough food for rabbits. We are interested in:

1. Determining whether the rabbits and foxes could co-exist in this environment.
2. Finding the equilibrium values of the system and determining if they are stable, unstable, or semistable.
3. Modeling the dynamics of the interaction between the predator foxes and the prey rabbits, so we can predict the populations of rabbits and foxes at any year.
4. Investigating the long-term behavior of the two species.

1.2.2. Step 2: Make Assumptions

It is very important to state the assumptions clearly. The construction of a mathematical model greatly depends on the assumptions. It is clear that a change in the assumptions would result in a different model. It is advisable to simplify assumptions, at least at the beginning, to make the model manageable. Therefore, some assumptions are made to simplify the model.

We will use the predator–prey example to illustrate step 2. Because the forest is a complex ecosystem, we need to make the following assumptions to simplify the predator–prey model:

i. There is enough food for the rabbits and the population of rabbits increases by a constant rate. That is, the rabbits' population increases exponentially.
ii. The population of rabbits decreases as a result of the interactions between rabbits and foxes.

iii. The rabbits are the only source of food for foxes. Therefore, in the absence of rabbits the population of decreases by a constant rate and dies out. That is, the foxes' population decreases exponentially.

iv. The population of foxes increases as a result of the interactions between rabbits and foxes.

v. The rabbits and foxes live in a closed environment. That is, that there is no interaction between these two species and other species, there is no emigration from or immigration to the forest, and there is no harvesting or hunting.

1.2.3. Step 3: Formulate the Mathematical Problem

In step 3 we enter the mathematics world. In the first part of this step we need to choose mathematical symbols for the variables and parameters. Recall that variables are quantities that change within the problem, whereas the parameters are constant within a problem.

For example, in the predator–prey example we might choose the following variables:

R_n = the population of the prey rabbits at the time period n.
F_n = the population of predator foxes at time period n.
n = the time periods in years, $n = 0, 1, 2, \ldots, k$.

For the parameters we might choose:

a = the natural growth rate of rabbits in the absence of foxes, and $a > 0$.
b = the death rate of rabbits as a result of the presence of foxes, and $b > 0$.
c = the natural decay rate of foxes in the absence of rabbits, and $c > 0$.
d = the growth factor of foxes due to the presence of rabbits, and $d>0$.

In the second part of step 3 we use the assumptions made in step 2 and the variables and parameters defined in the first part of step 3 to formulate the problem in mathematical notation. As a result of the formulation, the problem might be represented by a single algebraic, difference, differential, or matrix equation or by a system of algebraic, difference, or differential equations. The problem might be represented by an algorithm, and so on.

For example, the predator–prey model may be represented by a system of the following linear difference equations

$$R_{n+1} = R_n + aR_n - bF_n. \tag{1.21}$$

$$F_{n+1} = F_n - cF_n + dR_n. \tag{1.22}$$

or the following system of two nonlinear difference equations

$$R_{n+1} = R_n + aR_n - bR_nF_n. \tag{1.23}$$

$$F_{n+1} = F_n - cF_n + dR_nF_n. \tag{1.24}$$

It is necessary to use the selected representation to answer the posed questions for the real-world problem.

1.2.4. Step 4: Solve the Mathematical Problem (Model)

In step 4 we use appropriate available mathematical, computational, or graphical tools and techniques to solve the mathematical problem (model). The solution might be an analytical solution (a closed-form mathematical expression), a numerical solution, or a graph. It might also be the implementation of an algorithm or the running/testing of a simulation.

For example, there is an analytical solution of the predator–prey model represented by linear equations 1.21 and 1.22 in the form

$$X_n = T^n X_0,$$

where $T = \begin{bmatrix} 1+a & -b \\ d & 1-c \end{bmatrix}$, $X_n = \begin{bmatrix} R_n \\ F_n \end{bmatrix}$, and $X_0 = \begin{bmatrix} R_0 \\ F_0 \end{bmatrix}$ is the initial distribution vector.

The matrix algebra allows us to investigate the solution fully. However, the modeler will realize that the linear representations 1.21 and 1.22 of the predator–prey model are unrealistic.

The predator–prey model is a realistic one if it is represented by the nonlinear difference equations 1.23 and 1.24. Therefore, we will focus our attention on answering the questions of the real-world problems using the nonlinear representation. However, for nonlinear equations there is no analytical solution.

It can be easily shown that this system has two equilibrium values, say R_e and F_e, where

$$R_e = \frac{c}{d} \quad \text{and} \quad F_e = \frac{a}{b}.$$

Using MATLAB it can be concluded that these equilibrium values are unstable. With MATLAB we create two types of graphs: time series graphs (R_n and F_n vs. n) and phase plane graphs (F_n vs. R_n or R_n vs. F_n), for different values of the parameters a, b, c, and d and the initial values R_0 and F_0. Answers to the posed questions and the system's long-term behavior can be extracted from these graphs.

1.2.5. Step 5: Interpret the Solution

In step 5, the answers to the mathematical problem need to be interpreted in terms of the context of the real-world problem. The modeler must check the answers to ensure that the model answered the original real-world problem within the assumptions made in step 2 and the initial conditions.

In our predator–prey example the following might be interpretations to solutions obtained in step 4:

- The rabbits and foxes may co-exist together without extinction of one of the species. At certain values of the rabbits and foxes, there is no change in the species' populations. This interpretation is a conclusion of the existence of equilibrium values of rabbits and foxes.
- The equilibrium values of the rabbits and foxes are sensitive to small change. In other words, a small change to the equilibrium values makes the rabbits and foxes populations diverge from the equilibrium values. This interpretation is a conclusion from the instability of the equilibrium values established in step 4.

1.2.6. Step 6: Verify the Model

Next, it is necessary to verify the validity of the model. A common verification method is to compare the model's predicted results with known real-world data or with data obtained from an experiment designed to test the model. Be sure that all the necessary variables were used and all the assumptions were incorporated.

If the outcome of the verification is unsatisfactory, the modeler needs to refine the model, which is improving it. To refine the model you need to reexamine the modeling process starting with step 1. Be sure that you did not omit any necessary assumptions and variables. Check that the mathematical problem accurately represents the real-world problem. Check for the correctness of solving the mathematical problem.

If the verification of the model in step 6 is satisfactory, the modeler writes a report on the model, if he or she is required to submit one. The report should follow the requested format.

1.2.7. Exercises 1.2

In the following problems, real-world situations are briefly stated without specifics. For each situation do the following:

A. Formulate the real-world problem.

B. Make assumptions. If you made assumptions to simplify the problem, state them at the beginning.

C. Choose mathematical symbols for the variables and parameters.

D. Formulate the mathematical problem. Note that you are not asked to solve the problem.

1. The dynamics of the population of the United States of America.
2. The dynamics of the population of a single species.
3. The dynamics of the population of a single species, such as deer, with hunting.
4. The dynamics of the interaction of two species competing for the same food, such as foxes and wolves who compete for rabbits.
5. The dynamics of two interacting species, predator and prey, such as falcons and rats.
6. Obtaining the maximum sustainable yield of a natural renewable resource, such as a colony of whales.
7. Administering a drug, such as antibiotic, for a patient.
8. The dynamics of the spread of a contagious disease, such as flu, among the students of a college.

1.3. GETTING STARTED WITH MATLAB

In this section we introduce the basics for getting started with MATLAB. This introduction is by no means a reference or manual for the program. It is simply a quick start guide to writing simple commands, iterating difference equations, and producing graphs.

To start MATLAB, double click on the MATLAB icon on computer. You should see a screen broken into five parts (panels). The main part in the middle is the **Command Window**. On the left-hand side are two panels, **Current Folder** and **Details**. On the right-hand side are two panels, **Workspace** and **Command History**.

In the Command Window you enter the commands and get answers or error messages.

The MATLAB prompt is >>

Three main ways to do computations are

- Enter (type) the commands in the Command Window
- Use commands saved in a **script file**.
- Use a **function**, where the function code is saved in an **M-flie**.

1.3.1. Simple Arithmetic and Definition of Variables

The basic arithmetic operators are

+	addition
−	subtraction
*	multiplication
/	division
^	power operator

Examples

To set the variable x to be 3 type x=3 and hit enter key.

```
>> x=3
x=
  3
>> y=4;                    % The ; suppresses the output
>> x+y
ans=
  7
>> z=sqrt(x^2 + y^2)       % sqrt is the square root function
z=
  5
```

1.3.2. Vectors and Matrices

Vectors can be defined in two ways. The first method is to enter the elements of a vector in square brackets separated by spaces or by a comma (,):

```
>> v=[3 8 6 1]             % To enter the vector v=(3, 8, 6, 1)
v=
  3   8   6   1
>> u=[10, 12, 16]
u=
  10   12   16
```

The second method for defining vectors uses elements going in even steps from one value to another value

```
>> a=4:9                   % The vector a goes from 4 to 9 in increments of 1
a=
  4   5   6   7   8   9
>> b=5:3:22                % The vector b goes from 5 to 22 in increments of 3
b=
  5   8   11   14   17   20
>> c=1:0.5:4               % The vector c goes from 1 to 4 in increments of 0.5
c=
  1.0000 1.5000 2.0000 2.5000 3.0000 3.5000 4.000

>> v(1)                    % The first element of the vector v
ans=
  3
>> v(4)                    % The fourth element of the vector v
```

```
ans =
  1
>> v(5)=9                     % Additional elements can be added to a vector
v =
  3   8   6   1   9
>> b(2:5)                     % The subvector of b from the 2nd to 5th
                              % element
ans =
  8   11   14   17
>> x=[5 8 6]; y=[20 33];
>> z=[x y]                    % A new vector can be defined by previously
z =                           % defined vectors
  5   8   6   20   33
>> length(z)
ans =
  5
>> m=[3; 6; 9]                % To define column vector m = [3; 6; 9]
m =
  3
  6
  9
```

Matrices are defined by entering the rows separated by semicolons (;), and the entries of a row are separated by spaces or by a comma (,). The matrix entries are enclosed in square brackets.

For example matrices $A = \begin{bmatrix} 2 & 6 & 7 \\ 4 & 8 & 10 \end{bmatrix}$ and $B = \begin{bmatrix} -2 & 3 & 1 \\ 2 & -4 & -3 \end{bmatrix}$ may be entered as:

```
>> A=[2 6 7; 4 8 10]          % The elements of each row are separated
>>                            % by spaces
A =
  2 6  7
  4 8 10
>> B=[-2, 3, 1; 2, -4, -3]    % The elements of rows are separated by,
B =
  -2   3   1
   2  -4  -3
>> A(2, 3)                    % The element in row 2 and column 3 of matrix A
ans =
  10
>> A(2, :)                    % To extract the 2nd row of matrix A
```

```
ans =
  4 8 10
>> A(:, 2)                     % To extract the 2nd column of matrix A
ans =
  6
  8
>> A=[A; 1 3 5]                % To add the row [1 3 5] at the end of matrix A
A=
  2 6  7
  4 8 10
  1 3  5
>> B=[B [6; 8]]                % To add the column [6; 8] to matrix B
B=
  -2  3  1 6
   2 -4 -3 8
>>
```

1.3.3. Iteration

Let us consider the following situation. The population of a species increases every year by 10%. Letting y_n be the population at the end of n years and $y_1 = 100$ be the initial population, this situation is modeled by the recurrence relation (or difference equation)

$$y_{n+1} = 1.1y_n, \qquad y_1 = 100, \quad n = 1, 2, \ldots \tag{1.25}$$

To obtain the sequence $y_1, y_2, y_3, \ldots, y_7$, we use y_1 to get y_2, use y_2 to get $y_3, \ldots$, use y_6 to get y_7. To obtain the first 6 values of y_n after the initial value y_1 we use the difference equation (recurrence relation) 1.25 recursively. We have

$$\begin{aligned}
&\text{For } n = 1, \quad y_2 = 1.1y_1 = 1.1(100) = 110.\\
&\text{For } n = 2, \quad y_3 = 1.1y_2 = 1.1(110) = 121.\\
&\text{For } n = 3, \quad y_4 = 1.1y_3 = 1.1(121) = 133.10.\\
&\text{For } n = 4, \quad y_5 = 1.1y_4 = 1.1(133.10) = 146.41.\\
&\text{For } n = 5, \quad y_6 = 1.1y_5 = 1.1(146.41) = 161.051.\\
&\text{For } n = 6, \quad y_7 = 1.1y_6 = 1.1(161.051) = 177.1561.
\end{aligned}$$

This process is called *iteration* and can be obtained by a **for** loop in MATLAB.

One way to write code to iterate difference equation 1.25 is to create an array that contains the values of the initial value and the calculated values. Let us call the array y. It is a good practice to create the array with zeros—that is, an array with the required length and store zeros in it.

Here is one possible code (list of commands) for this task:

```
>> y=zeros(1, 7);          % Define y as an array of zeros with length 7 which
>>                         % will be used to store 7 terms of the sequence
>> y(1)=100;               % Store the initial value in the first slot of y,
>>                         % y(1)=100
>> for n=1:6               % This "for loop" starts at n=1 and ends
>>                         % at n=6
     y(n+1)=1.1*y(n);      % This is equation 1.25 and is the body of loop
   end;
>> y
y=
  100.0000 110.0000 121.0000 133.1000 146.4100 161.0510
  177.1561
```

Note that MATLAB considers anything written after a percent symbol (%) and to the end of the line as a comment and ignores it. You will need to write explanations and documentation for your programs in MATLAB.

Note that equation 1.25 can be written in the following form:

$$y_n = 1.1y_{n-1}, \qquad y_1 = 100, \quad n = 2, 3, \ldots \tag{1.26}$$

To obtain the first 6 values of y_n after the initial value y_1 we use the difference equation 1.26 recursively. We have

For $n = 2$, $y_2 = 1.1y_1 = 1.1(100) = 110.$

For $n = 3$, $y_3 = 1.1y_2 = 1.1(110) = 121.$

For $n = 4$, $y_4 = 1.1y_3 = 1.1(121) = 133.10.$

For $n = 5$, $y_5 = 1.1y_4 = 1.1(133.10) = 146.41.$

For $n = 6$, $y_6 = 1.1y_5 = 1.1(146.41) = 161.051.$

For $n = 7$, $y_7 = 1.1y_6 = 1.1(161.051) = 177.1561.$

Here is a code to do these calculations. Note the difference between the start and the end of the "for" loop in this and the previous code. Note that you obtain the same results.

```
>> y=zeros(1, 7);          % Define y as an array of zeros with length 7,
>>                         % which will be used to store 7 terms of the
>>                         % sequence
>> y(1)=100;               % Store the initial value in the first slot of y,
>>                         % y(1)=100
```

```
>> for n=2:7                 % This "for loop" starts at n=2 and ends at
>>                           % n=7
      y(n)=1.1*y(n-1);       % This is equation 1.26 and is the body of loop
   end;
>> y
y=
   100.0000 110.0000 121.0000 133.1000 146.4100 161.0510
   177.1561
```

1.3.4. Plotting

If y is an array of numbers of length k, the command `plot(y)` will plot $y(1), y(2), \ldots, y(k)$ verses x values: $1, 2, 3, \ldots, k$. For example,

```
>> x=1:10
x=
   1  2  3  4  5  6  7  8  9  10
>> y=[2 6 8 15 10 38 46 50 80 100]
y=
   2  6  8  15  10  38  46  50  80  100
>> plot (y, 'k*')
```

The graph from the command `plot(y, 'k*')` is shown in Figure 1.3A. MATLAB plots the discrete points (1, 2), (2, 6), (3, 8), (4, 15), …, (8, 50), (9, 80), (10, 100) in black stars (*).

Note that the command `plot(x, y, 'k*')` would produce the same graph as shown in Figure 1.3A.

The command `>> plot (x, y, 'k')` would produce the graph in Figure 1.3B.

The graph from the command `>> plot (x, y, 'k*', x, y, 'k')` is shown in Figure 1.3C, where the discrete points are in black stars (*), and these points are connected with segment lines in black.

Now let us assume that we have two data arrays x and y of the same length. The plot commands are as follows:

plot (x, y)	plots y vs. x and connects the dots
plot (x, y, 'b.')	plots discrete points of y vs. x in blue points (.), and the points are not connected
plot (x, y, 'r*', x, y, 'b')	plots discrete points of y vs. x in red stars (*), and the points are connected with segment lines in blue

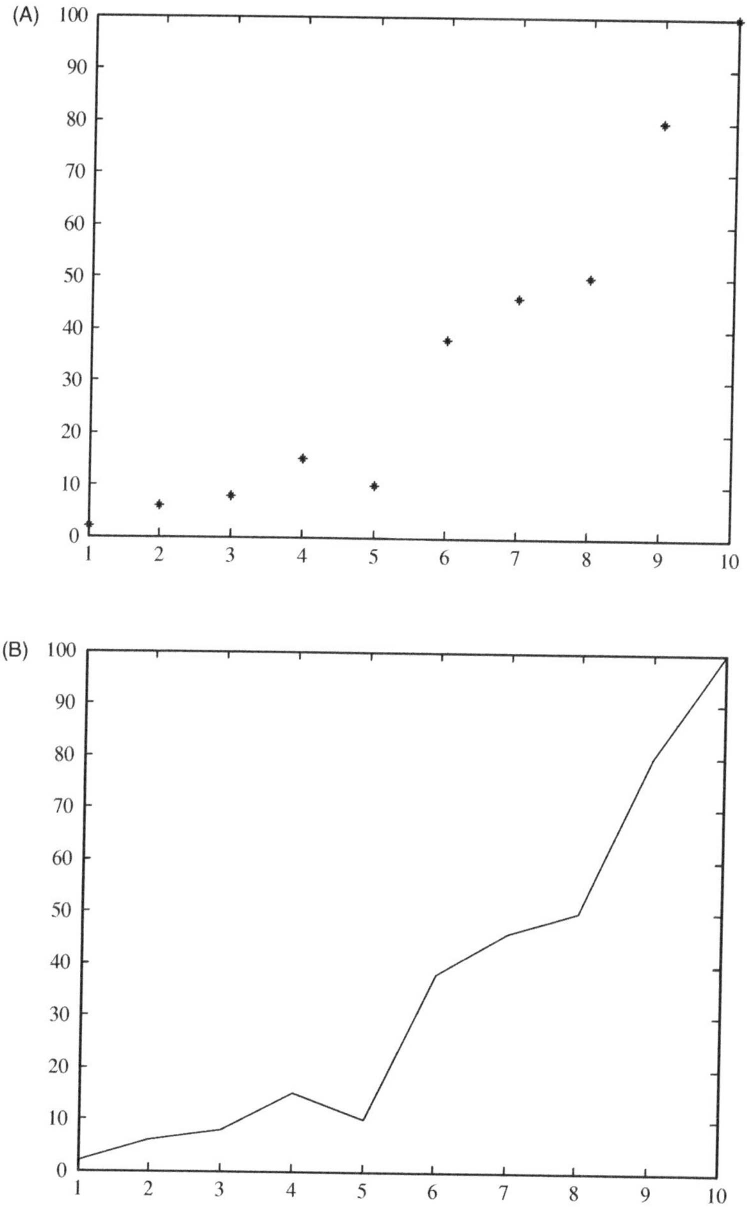

FIGURE 1.3. **A**. Graph from MATLAB command `plot(y, 'k*')` where **y** is an array of numbers, $y=[2, 6, 8, 15, 10, 38, 46, 50, 80, 100]$. **B**. Graph from MATLAB command `plot(x, y, 'k')` where x and y are arrays of numbers of the same length, with $x=[1, 2, 3, 4, 5, 6, 7, 8, 9, 10]$ and $y=[2, 6, 8, 15, 10, 38, 46, 50, 80, 100]$. **C**. Graph from MATLAB command `plot(x, y, 'k*', x, y, 'k')` where x and y are arrays of numbers of the same length, with $x=[1, 2, 3, 4, 5, 6, 7, 8, 9, 10]$ and $y=[2, 6, 8, 15, 10, 38, 46, 50, 80, 100]$.

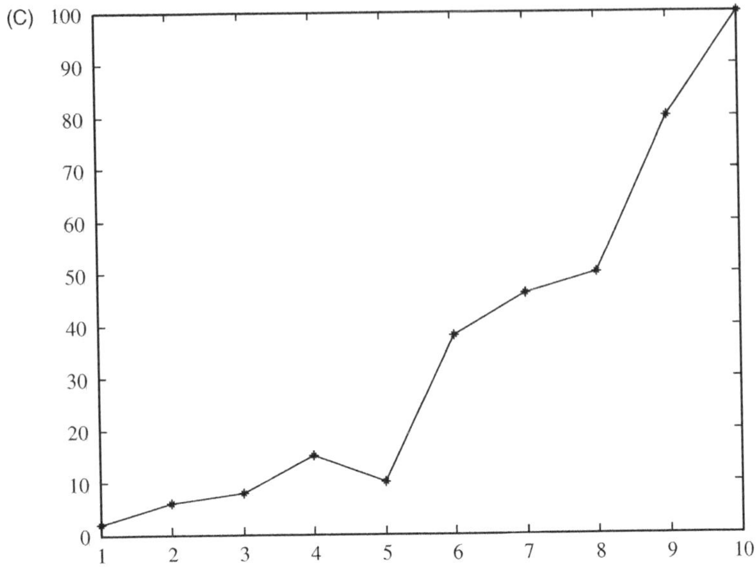

FIGURE 1.3. *(Continued)*

Example 1.1

Enter the following commands

```
>> x=-6:7
x=
  -6 -5 -4 -3 -2 -1 0 1 2 3 4 5 6 7
>> y1=x.^2
y1=
  36 25 16 9 4 1 4 9 16 25 36 49
>> y2=x.*2+8
y2=-4 -2 0 2 4 6 8 10 12 14 18 20 22
>> plot(x, y1, 'k.', x, y1, 'k')
```

The graph from the above plot command is shown in Figure 1.4A

If we want two graphs on one coordinate system, use the command **hold on**, which holds the current graph and the next graph will be plotted in the same window. For example, adding the following two commands to the earlier code will produce the graph in Figure 1.4B.

```
>> hold on
>> plot(x, y2, 'ko', x, y2, 'k')
```

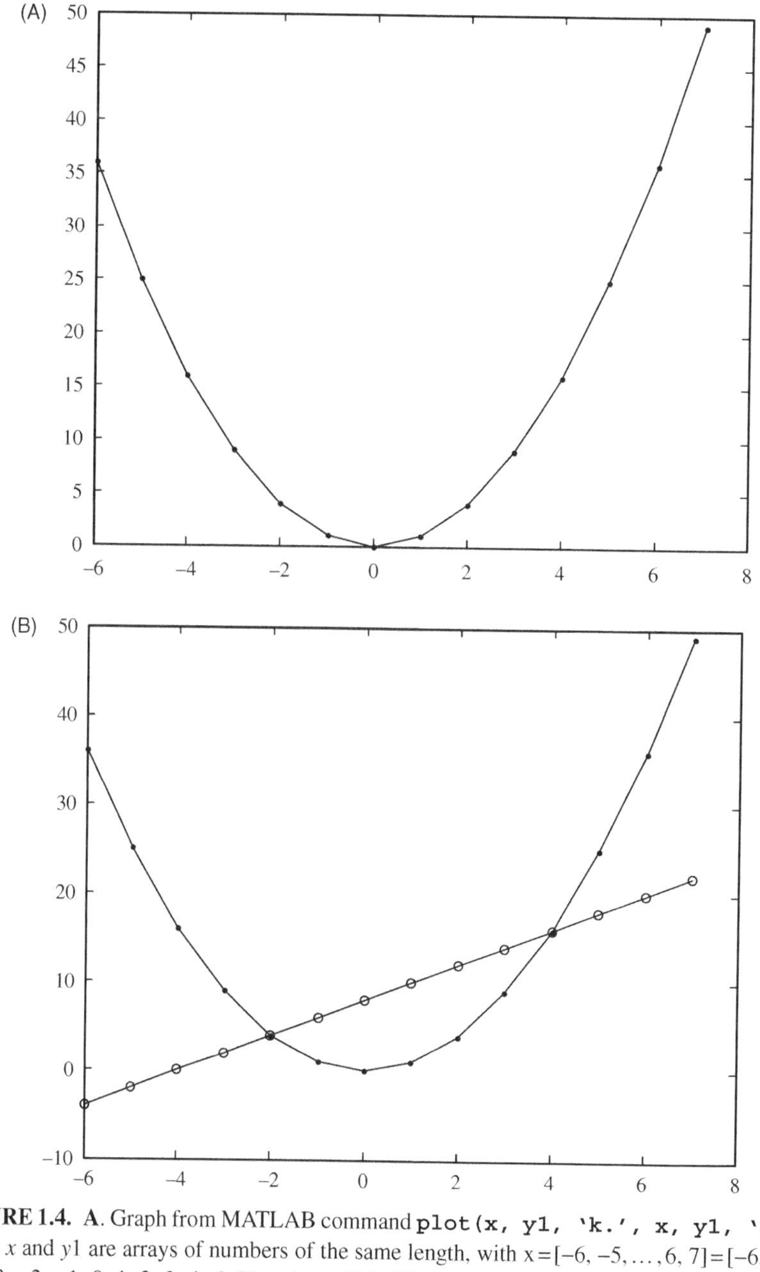

FIGURE 1.4. A. Graph from MATLAB command `plot(x, y1, 'k.', x, y1, 'k')` where x and $y1$ are arrays of numbers of the same length, with x = [–6, –5, …, 6, 7] = [–6, –5, –4, –3, –2, –1, 0, 1, 2, 3, 4, 6, 7] and $y1$ = [36, 25, 16, 9, 4, 1, 4, 9, 16, 25, 36, 49]. **B**. Two graphs on one coordinate system from MATLAB commands:

```
plot(x, y1, 'k.', x, y1, 'k')
hold on
plot(x, y2, 'ko', x, y2, 'k')
```

where x, $y1$, and $y2$ are arrays of numbers of the same length, with x = [–6, –5, –4, –3, –2, –1, 0, 1, 2, 3, 4, 6, 7], $y1$ = [36, 25, 16, 9, 4, 1, 4, 9, 16, 25, 36, 49], and $y2$ = [–4, –2, 0, 2, 4, 6, 8, 10, 12, 14, 18, 20, 22].

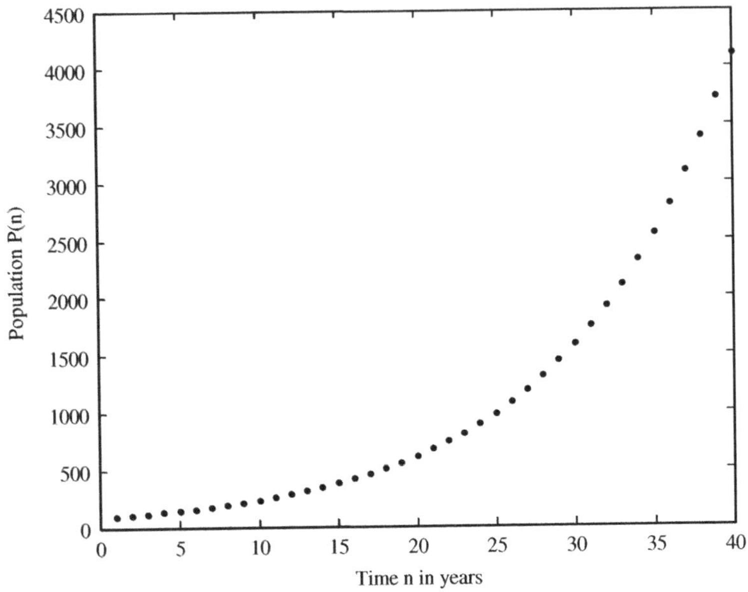

FIGURE 1.5. Graph of a species population, P_n, vs. the time, n, in years, with the labeling of the two axes, where $P_n = 1.1P_{n-1}$, $P_1 = 100$, and $n = 1, 2, \ldots, 40$.

Note that any other plots will be in the same window with other graphs as long as the command **`hold on`** is in effect. The command **`hold off`** will cancel **`hold on`** and plots a new graph in a new window.

Example 1.2

Consider the population of a species represented by the difference equation

$$P_n = 1.1P_{n-1}, \qquad P_1 = 100, \quad n = 2, 3, \ldots, 40,$$

Here is a possible code for iterating the difference equation, graph the population, p_n, vs. time, n, and label the two axes. The graph is shown in Figure 1.5.

```
>> T=1:40;                          % Define the time as an array
>> P=zeros(1, 40);                  % Define P as an array of zeros with length 40
>> P(1)=100;                        % Store the initial value in the first slot of P
>> for n=2:40                       % This "for loop" starts at n=2 and
>>                                  % ends at n=40
      P(n)=1.1*P(n-1);              % This is the iterated difference equation
   end;
>> plot(T, P, 'k.');                % Plots the population P(n) vs. time n in
>>                                  % black points
```

```
>> xlabel('Time n in years');     % Labels the x axis – time n
>> ylabel('Population P(n)');     % Labels the y axis – population P(n)
```

MATLAB uses the following symbols for color:

r	red
y	yellow
b	blue
c	cyan
k	black
m	magenta
g	green
w	white

MATLAB uses the following symbols for line styles

.	point
-	solid
*	star
:	dotted
o	circle
--	dashed
+	plus
-.	dash-dot
x	x-mark

For example, if $x1$ and $y1$ are data arrays of the same length, the command

```
>> plot(x1, y1, 'r*')
```

plots $y1$ vs. $x1$ in discrete red stars; and the command

```
>> plot(x1, y1, 'b--')
```

plots $y1$ vs. $x1$ in a discrete blue dashed line.

1.3.5. M-files

So far, we were working with commands entered into the Command Window. It is more efficient and convenient to enter a sequence of commands into a file, save it, and run the file. MATLAB allows one to create such files, which are called **M-files**.

An M-file is a collection of MATLAB commands saved in a file that has the extension ".m". An M-file is an ordinary text file and may be created with the MATLAB editor or with an ordinary text editor. To execute the commands in an M-file, type the name of the file (without the extension ".m") as a command in the MATLAB prompt (>>). The commands in the file will be executed one by one.

There are two categories of M-files: **script files** and **function files**. A script file is a collection of MATLAB commands. A function file is a user-defined function.

Script Files

We are interested in creating, saving, and executing an M-file. To illustrate the procedure of creating an M-file using the MATLAB Editor and executing the M-file, we will use Example 1.2 (p. 22). Recall that for the example, we entered the commands into the Command Window. The task is to iterate the difference equation that represents the population of a species

$$P_n = 1.1P_{n-1}, \qquad P_1 = 100, \quad n = 2, 3, \ldots, 40$$

and graph the population, P_n, vs. time, n, labeling the two axes.

1. On the menu bar at the top of the MATLAB screen click on **File**, select **New**, select **Script**. You have a blank screen (note that at the top of the screen you see **Editor – Untitled**)
2. Enter the following commands

```
T=1:40;                         % Define the time as an array of 40
                                % slots
P=zeros(1, 40);                 % Define P as an array of zeros
                                % with length 40
P(1)=100;                       % Store the initial value in the first
                                % slot of P
for k=2:40                      % This "for loop" starts at
                                % k=2 and ends at k=40
    P(k)=1.1*P(k - 1);          % This is the iterated difference
                                % equation
end;
plot(T, P, 'k.');               % Plots P(n) vs. n in discrete black
                                % dots
xlabel('Time n in years');      % Labels the x axis – time n
ylabel('Population P(n)');      % Labels the y axis – population
                                % P(n)
```

3. To save the file click on **File**, select **Save As**.... In the drive and folder of your choice enter the name of the file, say Population, without an extension. Assume that you created the folder MyMATLAB_MFiles on your C drive and you saved the file Population (without an extension) in that folder. You will see on the top screen

   ```
   Editor - C:\MyMATLAB_MFiles\Population.m
   ```

   Note that the MATLAB Editor has automatically extended the file name with .m.
4. To execute the file, click on the green icon on the top menu bar or in the Command Window and enter the following command

   ```
   >> Population
   ```

   If there are no mistakes in the contents of the script file, you will see a graph with the title Figure 1 on the MATLAB Graphing Window (this is the same graph as shown in Figure 1.5). However, if there are mistakes in the file, you will get error messages in the Command Window. You will then need to correct the errors and resave the file before executing it again.

Remarks on Script Files

Note that the variables in a script file are global. This means they are accessed in the workspace (Command Window). For example, the variables P and T in the script file Population can be accessed in the Command Window—for example, the population P_{10} is obtained by entering the following command

```
>> P(10)
ans =
   235.7948
```

It is informative to start an M-file with a comment that contains the file name plus a description of the script

```
% File name followed by description of the script file.
```

For example, let us add the following three lines to the code of the script file Population and save the file under the name **Population.m**:

```
% Population – A script file to iterate the difference equation
% representing the population of certain species, P(k)= 1.1P(k-1),
% P(1)=100, n=2, 3,...,40. The script file graphs P(n) vs. n.
```

```
T=1:40;                         % Define the time as an array of 40
                                % slots
P=zeros(1, 40);                 % Define P as an array of zeros with
                                % length 40
P(1)=100;                       % Store the initial value in the first slot
                                % of P
for k=2:40                      % This "for loop" that starts at k=2
                                % and ends at k=40
   P(k)=1.1*P(k - 1);           % This is the iterated difference equation
end;
plot(T, P, 'k.');               % Plots P(n) vs. time n in discrete black
                                % dots
xlabel('Time n in years');      % Labels the x axis – time n
ylabel('Population P(n)');      % Labels the y axis – population P(n)
```

If you enter the command **help Population** in the Command Window, you will get the description of the script file Population you entered in first comment lines after the file name. Here is the command and MATLAB output:

```
>> help Population
  Population - A script file to iterate the difference
  equation representing the population of certain species,
  P(k)=1.1P(k-1), P(1)=100, n=2,3,...,40. The script file
  graphs P(n) vs. n
```

Function Files

A function file is an M-file containing a user-defined function.

For example, let us write a function to calculate the volume, v, of rectangular box of length l, width w, and height h. We know that $v = l \cdot w \cdot h$. We will give a name to this function, say RecVolume. You could write the following commands in the MATLAB Editor:

```
function v=RecVolume(l, w, h)
% RecVolume is a function that calculates the volume of a rectangular box.
% The function input: l, w, and h, where l=length, w=width, and h=height.
% The function output is the volume v
v=l*w*h;
```

Save this code in a file under the name RecVolume.m. Note that if you use MATLAB Editor to write the function code and save the file, the extension .m will be automatically extended to the file name.

To call this function for specific values of l, w, and h, for example $l=10$, $w=20$, and $h=5$, we write the command

```
>> v=RecVolume(10, 20, 5)
v=
  1000
```

If $l=2\pi^2$, $w=\sqrt{124.26}$, and $h=3l$, the function RecVolume can be called as

```
>> RecVolume(2*pi^2, sqrt(124.26), 3*l)
ans=
  4.1476e+03
```

```
>> help RecVolume
  RecVolume is a function that calculates the volume of
  a rectangular box.
  The function input: l, w, and h, where l=length, w=width,
  and h=height.
  The function output is the volume v
```

Remarks on Function Files

1. The first line in the code of a function must be the keyword **function** followed by the output variable (or output variables) enclosed in square brackets, followed by the equals sign, followed by the input variables enclosed in parentheses and separated by comma. If the first line of a function file does not start with the keyword *function*, you will not be able to call the function and the file will be a script file.
2. The function descriptions in the comment lines are important. If the user enters the command **`help`** followed by function name, MATLAB will display the function description.
3. Variables used inside a function are local to the function (in other words, these variables cannot be called globally, in the command window).
4. The list of output variables of a function may be omitted if you are not interested in their values. For example if you write a function that iterates a difference equation and graphs the ordered pairs, but you are not interested in values of the ordered pairs, you may omit an output variable.

CHAPTER 2

MODELING WITH FIRST-ORDER DIFFERENCE EQUATIONS

2.1. MODELING WITH FIRST-ORDER LINEAR HOMOGENOUS DIFFERENCE EQUATIONS WITH CONSTANT COEFFICIENTS

In this section, we investigate situations that are modeled by first-order linear homogenous difference equations with constant coefficients, that is, equations in the form

$$y_{n+1} = ay_n, \tag{2.1}$$

where a is a constant coefficient. We will iterate equation 2.1 with an initial condition to find a numerical solution, and we will also derive an analytical solution of equation 2.1.

2.1.1. Model 2.1: Drugs

Assume that the kidneys remove 20% of a drug from the blood every 4h. Assume that the initial dose of the drug is 200mg. Let d_n denote the amount of drug in the blood after n 4-h periods, and d_0 denote the initial amount of drug in the blood.

i. Find a difference equation that represents this situation.
ii. Find the amount of drug after 12 hours.
iii. Iterate the difference equation obtained in part I with the initial condition to find the ordered pairs (n, d_n), $n = 1, 2, \ldots, 28$. Graph it.

Explorations of Mathematical Models in Biology with MATLAB®, First Edition. Mazen Shahin.

iv. Find an analytical solution of the obtained difference equation. Use this solution to find the amount of drug in the blood after 1 day.

v. When will the amount of the drug reach 1 mg?

Discussion

i. The amount of drug in the blood after $(n+1)$ 4-h periods, d_{n+1}, equals the amount of drug after n 4-h periods, d_n, minus 20% of d_n. We obtain

$$d_{n+1} = d_n - 20\% \quad \text{of} \quad d_n = d_n - 0.2d_n = 0.8d_n.$$

Therefore this situation is modeled by the difference equation

$$d_{n+1} = 0.8d_n, \tag{2.2}$$

which is a first-order linear homogenous difference equation with constant coefficients.

ii. There are three 4-h periods in 12 h. So the amount of drug in the blood after 12 h is d_3. To find d_3, we will iterate the difference equation 2.2 with the initial condition

$$d_0 = 200. \tag{2.3}$$

To find d_1 put $n=0$ in equation 2.2 and use equations 2.3,

$$d_1 = 0.8d_0 = 0.8(200) = 160.$$

Similarly,

$$d_2 = 0.8d_1 = 0.8(160) = 128,$$

$$d_3 = 0.8d_2 = 0.8(128) = 102.4.$$

iii. The following MATLAB code can be used to create the table (n, d_n), $n = 0, 1, 2, \ldots, 30$. Note that this table is 31×2 matrix, where the first column is the time n and the second column is the amount of drug d_n.

```
t=0;
d=200;
M=[t d];
for k=1:30
    d=0.8*d;
    M=[M; k d];
end;
```

A partial output of the matrix M is

0	200.0000
1.0000	160.0000
2.0000	128.0000
3.0000	102.4000
4.0000	81.9200
5.0000	65.5360
....	....
25.0000	0.7556
26.0000	0.6045
27.0000	0.4836
28.0000	0.3869
29.0000	0.3095
30.0000	0.2476

To graph d_n vs. n we need to graph the second column of M vs. the first column. You may enter the code to have the graph in black points and label the axes.

```
plot(M(: , 1), M(: , 2), 'k.');
xlabel ('Time n in four-hour periods');
ylabel ('Amount of drug d(n)')
```

The graph is shown in Figure 2.1A. If you want to have the graph in black points and connect these points with black line graph, you may replace the above plot command with

```
plot(M(: , 1), M(: , 2), 'k.', M(: , 1), M(: , 2), 'k');
```

The graph is shown in Figure 2.1B.

iv. We will derive a closed-form of the solution (analytical solution) of the difference equation 2.2. We obtain

$$d_1 = 0.8d_0.$$
$$d_2 = 0.8d_1 = 0.8(0.8d_0) = (0.8)^2 d_0.$$
$$d_3 = 0.8d_2 = 0.8(0.8)^2 d_0 = (0.8)^3 d_0.$$

Following this pattern, we can conclude that

$$d_n = (0.8)^n d_0. \tag{2.4}$$

Equation 2.4 is the analytical solution of the difference equation 2.2. Because there are six 4-h periods in one day, the amount of the drug in the blood after

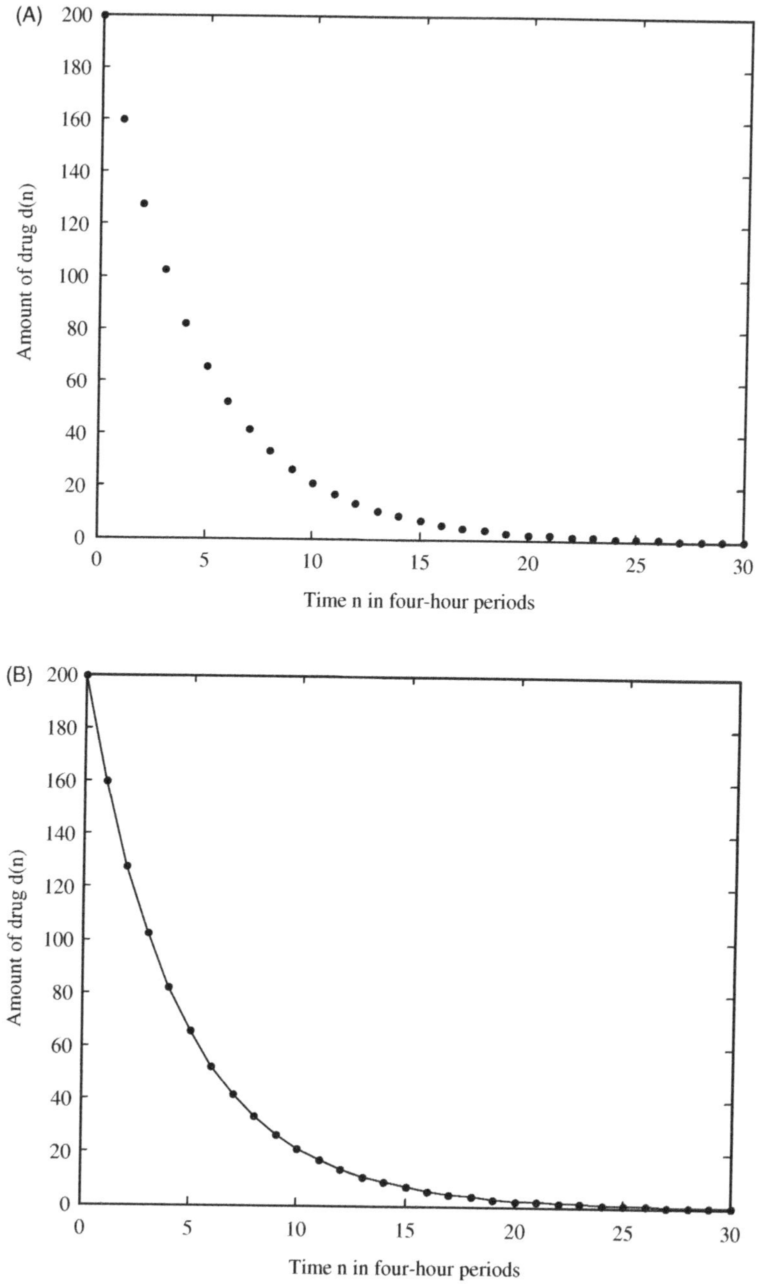

FIGURE 2.1. Graphs of the amount of drug in the blood after n time periods, d_n, vs. time, n, in 4-h periods. The graphs are in disconnected black points, and the axes are labeled.

1 day, d_6, can be calculated from equation 4.4 by setting $n=6$ and $d_0=200$. We obtain d_6 for the nearest hundredth

$$d_6 = (0.8)^6\,200 = 52.43.$$

Thus the amount of the drug in the blood after 1 day is 52.43 mg.

v. We want to know the value of n when $d_n=1$. Setting $d_n=1$ in equation 4.4, we get

$$1 = (0.8)^n\,200.$$

$$\frac{1}{200} = 0.8^n.$$

To solve this exponential equation, compute

$$\log\left(\frac{1}{200}\right) = \log(0.8^n).$$

$$\log\left(\frac{1}{200}\right) = n\log(0.8).$$

$$n = \frac{\log(1/200)}{\log(0.8)} = 23.7439.$$

Hence $n=23.7439$. This means that the amount of drug in the blood reaches 1 mg after approximately 24 4-h periods—that is, after 96 h, or 4 days.

Analytical Solution

The analytical solution of the first-order linear homogenous difference equation 2.1,

$$y_{n+1} = ay_n$$

can be derived in a similar way to model 2.1. Setting $n=0$ in equation 2.1, we obtain

$$y_1 = ay_0.$$

Put $n=1$ in equation 2.1 to obtain

$$y_2 = ay_1 = a(ay_0) = a^2 y_0.$$

Setting $n=2$ in equation 2.1, we get

$$y_3 = ay_2 = a(a^2 y_0) = a^3 y_0.$$

From this pattern, we conclude that

$$y_n = a^n y_0. \tag{2.5}$$

Solution of a First-Order Linear Homogenous Difference Equation with Constant Coefficients

An **analytical solution** to the difference equation 2.1

$$y_{n+1} = ay_n,$$

where a is a constant, is a formula that expresses y_n in terms of n, and the constants y_0 and a. That is, y_n as a function of n. The analytical solution of equation 2.1 is equation 2.5:

$$y_n = a^n y_0,$$

where y_0 is the initial condition.

Analytically we can check that equation 2.5 is a solution of equation 2.1 by showing that y_n in equation 2.5 satisfies equation 2.1. We have

$$\text{LHS of equation 2.1} = y_{n+1} = a^{n+1} y_0.$$

$$\text{RHS of equation 2.1} = ay_n = a(a^n y_0) = a^{n+1} y_0.$$

Therefore, LHS = RHS.

2.1.2. Model 2.2: Population Dynamics, First Pass

A population of owls is growing at 4% per year and there are 1000 owls now. For simplicity we will ignore the interaction of owls with other species. Let P_n be the population of owls n years from now.

i. Model the owls' population dynamic by a difference equation, which is a difference equation describing the change of the population year after year.
ii. Find the ordered pairs (n, P_n), $n = 0, 1, 2, \ldots, 60$. Graph P_n vs. n.
iii. Find an analytical solution of the difference equation obtained in a part ii. Use this solution to determine the population of owls after 10, 40, and 55 years.
iv. When does the population of owls double and triple?
v. What is the long-term behavior of the owls' population?

Discussion

i. The population of owls can be modeled by the difference equation

$$P_{n+1} = P_n + 0.04P_n.$$

$$P_{n+1} = 1.04P_n, \quad P_0 = 1,00(\qquad (2.6)$$

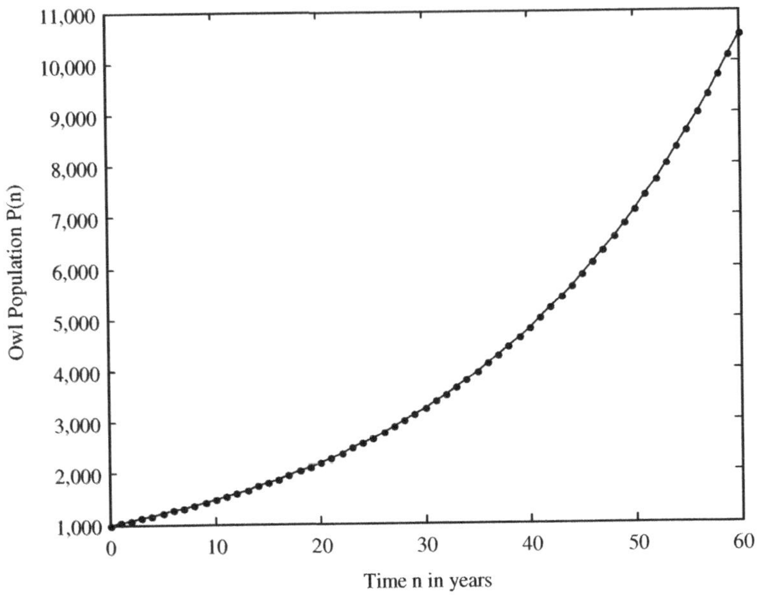

FIGURE 2.2. Graph of owls' population, P_n, vs. the time, n, in years. The initial population is $P_0 = 1{,}000$.

ii. The following MATLAB code may be used to find the ordered pairs (n, P_n), $n = 0, 1, 2, \ldots, 60$ and graph P_n vs. n, which is shown in Figure 2.2.

```
t=0;
P=1000;
M=[t P];
for k=1:60
    P=1.04*P;
    M=[M; k P];
end;
plot(M(: , 1), M(: , 2), 'k.', M(: , 1), M(: , 2), 'k');
xlabel ('Time n in years');
ylabel ('Population P(n)');
```

iii. Using equation 2.5, the analytical solution of the difference equation 2.6 is

$$P_n = (1.04)^n 1000 \tag{2.7}$$

Substitute $n = 10$, $n = 40$ and $n = 55$ in equation 2.7 to find P_{10}, P_{40} and P_{55}, respectively.

$$P_{10} = (1.04)^{10} 1000 = 1{,}480.24 \approx 1{,}480.$$

$$P_{40} = (1.04)^{40} 1000 = 4801.02 \approx 4801.$$

$$P_{55} = (1.04)^{55} 1000 = 8646.36 \approx 8646.$$

iv. We will use equation 2.7 to determine the value of n which satisfies $P_n = 2{,}000$. We have,

$$2000 = (1.04)^n 1000.$$
$$2 = (1.04)^n.$$

To solve for n we take log of both sides of this equation:

$$\log 2 = \log 1.04^n.$$
$$\log 2 = n \log 1.04.$$
$$n = \frac{\log 2}{\log 1.04} = 17.6729 \approx 18 \text{ years}.$$

That, is the owl's population will double in approximately 18 years. Similarly, the population will triple in 28 years. These calculations can be determined from the table obtained in part ii.

v. From the table (n, P_n), $n = 0, 1, 2, \ldots, 60$ and its graph (Figure 2.2), it is clear that the population of owls increases without bound. That is

$$\lim_{n \to \infty} P_n = \infty$$

2.1.3. Radioactive Decay

A radioactive substance is unstable, and due to chemical nuclear changes, it transforms into another substance. It loses its radioactivity over time by a fixed percentage. This process is called radioactive decay. The measure for decay rate of substance is called the half-life. The half-life of a radioactive substance is the time it takes for one-half of active atoms to decay. For example, iodine-131 has a half-life of 8 days. This means that if we have a 1,000 g specimen of iodine-131, 500 g will be left after 8 days (500 g are converted into some other element at the end of 8 days). After another 8 days, 250 g will be left, and after another 8 days, 125 g will be left, and so on. Note that the amount of substance will never be zero, but it will eventually be very small and can therefore be ignored.

Note that the half-life of a radioactive substance does not depend on the initial amount or the present amount of a substance. Each radioactive substance has its own half-life. Some radioactive substances have a half-life of seconds, hours, days, years, and billions of years. The half-life of some radioactive substances is shown in Table 2.1.

TABLE 2.1 Radioactive Substance Half-life

Krypton-91	10 s
Cobalt-55	18.2 h
Iodine-124	4.5 d
Iodine-131	8.0 d
Cobalt-61	5.3 d
Plutonium-241	13 yr
Cobalt-14	5,770 yr
Plutonium-239	24,400 yr
Uranium-238	4.5 billion yr

Model 2.3: Radioactive Decay

A hospital purchased 1,000 g of iodinc-131 to be used in their research, such as in locating brain tumors. The half-life of iodine-131 is 8 days. Let y_n be the amount of iodine-131 after n days and k be the decay constant/day.

i. Represent this situation by a difference equation—that is, find a relationship between y_{n+1} and y_n.
ii. Find the decay constant/day k.
iii. What is the amount of iodine-131 after 6 weeks?
iv. When will the amount of iodine-131 become 1 g?

Discussion

i. The relationship between y_n and y_{n-1} can be represented by the difference equation

$$y_{n+1} = y_n - ky_n.$$
$$y_{n+1} = (1-k)y_n.$$

ii. The analytical solution of this difference equation is

$$y_n = (1-k)^n y_0.$$

Because the half-life of iodine-131 is 8 days, we have

$$y_8 = 0.5y_0.$$

We will substitute $n=8$ in the analytical solution,

$$y_8 = (1-k)^8 y_0.$$

Therefore,

$$(1-k)^8 y_0 = 0.5 y_0.$$
$$(1-k)^8 = 0.5.$$
$$(1-k) = 0.5^{\frac{1}{8}} = 0.917004.$$
$$k = 1 - 0.917004 = 0.082996.$$

The analytical solution of the difference equation is

$$y_n = (1-k)^n y_0 = (0.917004)^n 1000.$$

iii. The amount of iodine-131 after 6 weeks (42 days) is y_{42},

$$y_{42} = (0.917004)^{42} 1000 = 26.2779\, g.$$

iv. We are looking for n such that $y_n = 1$. Put $y_n = 1$ in the analytical solution and solve for n. We have

$$1 = (0.917004)^n 1000.$$
$$\frac{1}{1000} = (0.917004)^n.$$
$$\log(0.001) = \log(0.917004)^n.$$
$$\log(0.001) = n\log(0.917004).$$
$$-3 = n(-0.0376287).$$
$$n = -3/(-0.0376287) = 79.7263 \approx 80.$$

Therefore, the amount of iodine-131 becomes 1 g after approximately 80 days.

2.1.4. Carbon Dating

The following principle is used by archeologists to estimate the age of once-living archaeological findings, such as bone or wood. There are two isotopes of carbon: radioactive carbon-14 and nonradioactive carbon-12. Carbon-14 and carbon-12 are absorbed in small amounts by all living tissues (animal and plant). As long as an animal or plant is alive, the ratio of carbon-14 to carbon-12 is a fixed constant and is the same as in the atmosphere. When an organism dies, no new carbon-14 is absorbed by the organism, and the existing carbon-14 slowly decays. The half-life of carbon-14 is 5730 years. The ratio of carbon-14 to carbon-12 in a fossil is measured and can be used to estimate the age of the fossil (i.e., the date when the organism died). Note that the ratio of carbon-14 to carbon-12 in the atmosphere is constant over the millennia.

This principle was developed in 1946 by Willard Libby, who received a Nobel Prize in chemistry.

Model 2.4: Carbon Dating

The laboratory testing revealed that 30% of carbon-14 is missing from a bone fragment. Knowing that the half-life of carbon-14 is 5730 years, how old is the bone?

Discussion

Assume that the carbon-14 decays at a constant rate k per year. Let y_n be the amount of carbon-14 in the bone after n years, where n is measured from the death year of the animal. y_0 is the initial amount of carbon-14 in the bone. We obtain

$$y_{n+1} = y_n - ky_n.$$

$$y_{n+1} = (1-k)y_n.$$

This is a first-order linear homogenous difference equation with constant coefficients. The analytical solution of the difference equation is

$$y_n = (1-k)^k y_0.$$

To determine k we will use the formula for the analytical solution and the fact that the half-life of carbon-14 is 5730 years:

$$y_{5730} = 0.5y_0$$

We have

$$y_{5730} = (1-k)^{5730} y_0.$$

We equate the right hand sides of these equations

$$0.5y_0 = (1-k)^{5730} y_0.$$

$$0.5 = (1-k)^{5730}.$$

$$(0.5)^{\frac{1}{5730}} = 1-k.$$

$$0.999477 = 1-k.$$

Hence

$$k = 0.000523.$$

The analytical solution is

$$y_n = (1-k)^n y_0 = (0.999477)^n y_0.$$

Because 30% of carbon-14 is missing, the amount of carbon-14 now, y_n, is

$$y_n = 0.70 y_0.$$

Therefore

$$(0.999477)^n y_0 = 0.7 y_0.$$

To solve for n take the log of both sides of the equation:

$$\log(0.999477)^n = \log 0.7.$$
$$n \log(0.999477) = \log(0.7).$$
$$n(-0.000523136) = -0.356674.$$
$$n = \frac{-0.356674}{-0.000523136}.$$
$$n \approx 681.799.$$

Therefore, the bone is approximately 682 years old.

2.1.5. Exercises 2.1

1. For the following difference equations figure out by hand y_1 through y_4.

A. $y_{n+1} = 2y_n,\ y_0 = 5$
B. $y_{n+1} = 0.9y_n,\ y_0 = 100$
C. $y_{n+1} = -2y_n,\ y_0 = -10$
D. $y_{n+1} - 3y_n = 0,\ y_0 = 4$

In Exercises 2–6,

A. Calculate, without computer, the first four terms after the initial condition, y_1, y_2, y_3, and y_4.
B. Find the analytical solution of the difference equation and use it to find y_4, y_{10}, and y_{20}.
C. Use MATLAB to compute and graph the ordered pairs (n, y_n), $n = 0, 1, 2, \ldots, 20$.

2. $y_{n+1} = 1.05y_n,\ y_0 = 120.$

3. $y_{n+1} = 0.9y_n,\ y_0 = 240.$

4. $y_{n+1} = 1.25y_n,\ y_0 = 60.$

5. $2y_{n+1} - 2.04y_n = 0,\ y_0 = 200.$

6. $y_{n+1} + 0.8y_n = 0,\ y_0 = -400.$

7. Assume the kidneys remove 25% of a drug from the blood every 4 h. Assume that the initial dose of the drug is 300 mg. Let D_n denote the amount of drug in the blood after n 4-h periods, and D_0 denote the initial amount of drug in the blood.

A. Find the difference equation that models this situation.

B. Find the amount of drug in the blood after 16 h.

C. Iterate the difference equation obtained in A with the initial condition to find the numerical solution (n, D_n), $n = 0, 1, 2, \ldots, 24$. Graph it.

D. Find an analytical solution of the obtained difference equation. Use this solution to find the amount of drug in the blood after 1 day

E. When will the amount of the drug reach 5 mg?

In Exercises 8–12, the average percent change (growth rate) is defined to be: (Number of births − Number of deaths) per 100 persons + (Number of immigrants − Number of emigrants) per 100 births.

8. The average percent change in the United States in years 2001–2010 was 0.918%. The estimated U.S. population in 2010 was 310,232,900.

A. Assuming that this trend continues, estimate the population in 2011, 2015, 2020, and 2030.

B. What was the approximate population in 2005?

9. The average population growth rate of Bulgaria in 2001–2010 was −0.502%, one of the lowest in the continent of Europe. The population of Bulgaria in 2010 was 7,403,000. At this rate what will the population be in 2015 and 2020?

10. The population of Norway in 2009 was 4,888,800. The 2005–2010 official estimation of the population growth was low and was estimated to be 0.62% per year during the mid-2000s. What was the population in the year 2012?

11. The population of Egypt in years 2005 and 2009 was 77,505,756 and 78,645,000, respectively.

A. Find Egypt's population growth rate.

B. Predict the population of Egypt in the year 2015

C. At this growth rate, when will the population of Egypt double?

12. The UN estimates the world population reached 4.8 billion in 1985, and the growth rate increase is 1.7% per year.

A. Estimate the world population in 1995, 2000, and 2009. (The actual world population in 1995, 2000, and 2009 was about 5.32 million, 6 billion, and 6.83 billion, respectively)

B. When will the world population double?

C. If 1 acre of land provides food for one person and the world has 8 billion acres of land, estimate when the arable land will not be enough to support food for the population.

13. The radioactive substance cobalt-60 is usually used in hospitals as radiation therapy to halt (interrupt) the development of cancer. The half-life of cobalt-60 is 5.3 years. The lab at General Hospital received 100 g of cobalt-60 on June 1, 2010.

A. Represent this situation by a difference equation.

B. Use the difference equation to determine the remaining amount of cobalt-60 on June 1, 2014.

C. When will the remaining amount of cobalt-60 be 5 g?

D. Use this model to determine the remaining amount of cobalt-60 on August 1, 2016. On January 1, 2020.

14. The research lab at Mercy Hospital has 150 g of iodine-131.

A. Determine the amount of the substance remaining after 3 weeks, knowing that the half-life of iodine-131 is 8 days.

B. When will the remaining amount of iodine-131 be 1 g?

15. A physicist received 6.4 g of plutonium-241 to be used in her experiment. After how many years will the remaining amount be 0.5 g? Note that the half-life of plutonium-241 is 13 years.

16. The radioactive substance cobalt-55, used in the diagnosis of certain medical ailments, has a half-life of 18.2 h. If a patient is given 40 mg, how many milligrams will be remaining after

A. 2 h?

B. 2 days?

C. 1 week?

17. The radioactive substance plutonium-238, used as a source of electrical power, has a half-life of 86 years. If a power cell in a spacecraft contains 2 g of plutonium-238, how many grams will be remaining after:

A. 5 days?

B. 3 weeks?

C. 4 months?

D. 10 years?

18. Laboratory testing reveals that 45% of carbon-14 is missing from an Egyptian mummy. How old is the mummy? Note that the half-life of carbon-14 is 5,730 years.

19. A chemical analysis showed that an ancient coffin had 40% of carbon-14 present in the wood. How old is the coffin? Note that the half-life of carbon-14 is 5,730 years.

20. Laboratory testing on a charcoal sample formed from a tree and collected from a volcanic area in Pompey, Italy, showed that the charcoal had 80% of carbon-14 present. When did the volcano erupt in Pompey?

2.2. MODELING WITH NONHOMOGENOUS FIRST-ORDER LINEAR DIFFERENCE EQUATIONS

In this section, I investigate some mathematical models that are represented by first-order linear nonhomogenous difference equations, that is equations in the form

$$y_{n+1} = a_n y_n + b_n, \quad n = 0, 1, 2, \ldots, \tag{2.8}$$

where a_n and b_n are functions of n or constants. We will focus our investigations on modeling with first-order linear nonhomogenous difference equations with constant coefficients in the form

$$y_{n+1} = a y_n + b, \tag{2.9}$$

where a and b are constants. We will iterate equations 2.8 and 2.9 with initial conditions to find numerical solutions, and we will also derive an analytical solution of equation 2.9.

2.2.1. Model 2.5: Drugs revisited

Suppose that the kidneys remove 20% of a drug from the blood every 4 h. Assume that a patient takes an initial dose of the drug followed by a dose of 20 mg of the same drug every 4 h. Let d_n denote the amount of drug in the blood at the end of n 4-h periods, and d_0 denote the initial amount of drug in the blood—that is, the initial dose of the drug. For each of the following initial values:

i. $d_0 = 140$ mg,
ii. $d_0 = 70$ mg,
iii. $d_0 = 100$ mg,

do the following:

A. Find a difference equation that models the situation. Determine the amount of drug in the blood after 20 h and 2 days.
B. Find a numerical solution (n, d_n), $n = 0, 1, 2, \ldots, 30$. Graph this solution and describe the graph.

Discussion

i. A. We want to find a relationship between d_{n+1} and d_n in the interval $(n, n+1)$. Because at the end of the $(n+1)$ st 4-h period, 20% of d_n is removed and 20 mg of drug is added, we have

$$d_{n+1} = d_n - 20\% \quad \text{of} \quad d_n + 20 = d_n - 0.2d_n + 20 = 0.8d_n + 20.$$

Therefore this situation is modeled by the difference equation and initial condition 2.10,

$$d_{n+1} = 0.8d_n + 20, \quad d_0 = 140. \tag{2.10}$$

Equation 2.10 is a first-order linear nonhomogenous difference equation with constant coefficients.

Because there are five 4-h periods in 20 h, we want to find d_5, which can be found by hand or by technology iteration. We have

$$d_1 = 0.8d_0 + 20 = 0.8(140) + 20 = 132.$$
$$d_2 = 0.8d_1 + 20 = 0.8(132) + 20 = 125.6.$$
$$d_3 = 0.8d_2 + 20 = 0.8(125.6) + 20 = 120.48.$$
$$d_4 = 0.8d_3 + 20 = 0.8(120.48) + 20 = 116.38.$$
$$d_5 = 0.8d_4 + 20 = 0.8(116.38) + 20 = 113.11.$$

Thus the amount of the drug in the blood after 20 h is 113.11 mg.

B. One of the following MATLAB codes may be used to find the numerical solution (n, d_n), $n = 0, 1, 2, \ldots, 30$ and graph it.

Version 1

```
t=0;
d=140;
M=[t d];
for k=1:30
    d=0.8*d+20;
    M=[M; k d];
end;
plot(M(: , 1), M(: , 2), 'k.', M(: , 1), M(: , 2),
'k');
xlabel ('Time n in 4-hour units');
ylabel ('Amount of drug d(n) in mg');
```

Version 2

```
t=0:30;                 % Create array of 31 slots for time
d=zeros(1, 31);         % Create 31-slot array d of zeros for
                        % drug amounts
d(1)=140;               % Store the initial amount of drug in the
                        % first slot
                        % Note that d0=140 is stored in the d(1)

for k=1:30
    d(k+1)=0.8*d(k)+20;
end;
plot (t, d, 'k.', t, d, 'k');
xlabel ('Time n in 4-hour units');
ylabel ('Amount of drug d(n) in mg');
```

The graph of these ordered pairs is shown in Figure 2.3A. As you can see, the amount of the drug in the blood at the end of each interval is smaller than the amount of the drug in the blood at the end of the previous interval. Eventually the amount of the drug in the blood reaches 100 mg and will remain without changing. Why?

ii. A. This situation is the same as in situation i, but the initial dose of the drug is 70 mg. Consequently, this situation is modeled by the difference equation and initial condition 2.11,

$$d_{n+1} = 0.8d_n + 20, \quad d_0 = 70, \tag{2.11}$$

Iterating equation 2.11, we get

$$d_1 = 0.8d_0 + 20 = 0.8(70) + 20 = 76,$$

$$d_2 = 90.8, \quad d_3 = 84.64, \quad d_4 = 87.71, \quad \text{and} \quad d_5 = 90.17.$$

Thus the amount of the drug in the blood after 20 h is 90.17 mg.

ii. B. Similar to situation i, the MATLAB code given earlier with $d_0 = 70$ produces the graph in Figure 2.3B. The graph shows that the amount of the drug in the blood increases, but it approaches 100 mg and levels off. In other words, the drug level approaches a limiting value of 100 and stays constant without change. Explain why.

iii. A. The third situation is modeled by the difference equation with initial condition 2.12,

$$d_{n+1} = 0.8d_n + 20, \quad d_0 = 100. \tag{2.12}$$

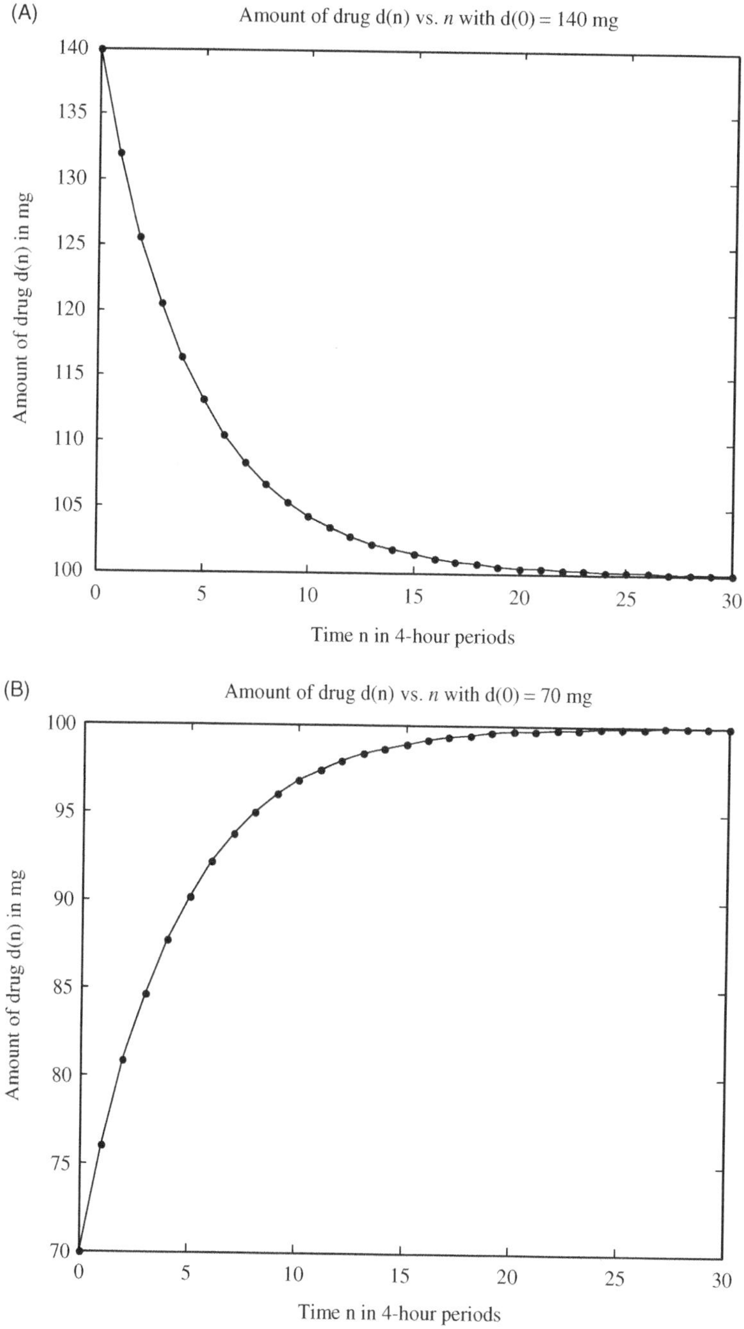

FIGURE 2.3. Graphs of the amount of drug in the blood, d_n, vs. the time, n, in 4-h periods. **A**. The initial value $d_0=140$ mg and $n=0,1,\ldots,30$, where $d_{n+1}=0.8d_n+20$. **B**. The initial value $d_0=70$ mg and $n=0,1,\ldots,30$, where $d_{n+1}=0.8d_n+20$. **C**. The initial value $d_0=100$ mg and $n=0,1,\ldots,30$, where $d_{n+1}=0.8d_n+20$. **D**. The initial values $d_0=70$ mg, $d_0=100$ mg, and $d_0=140$ mg, and $n=0,1,\ldots,30$, where $d_{n+1}=0.8d_n+20$.

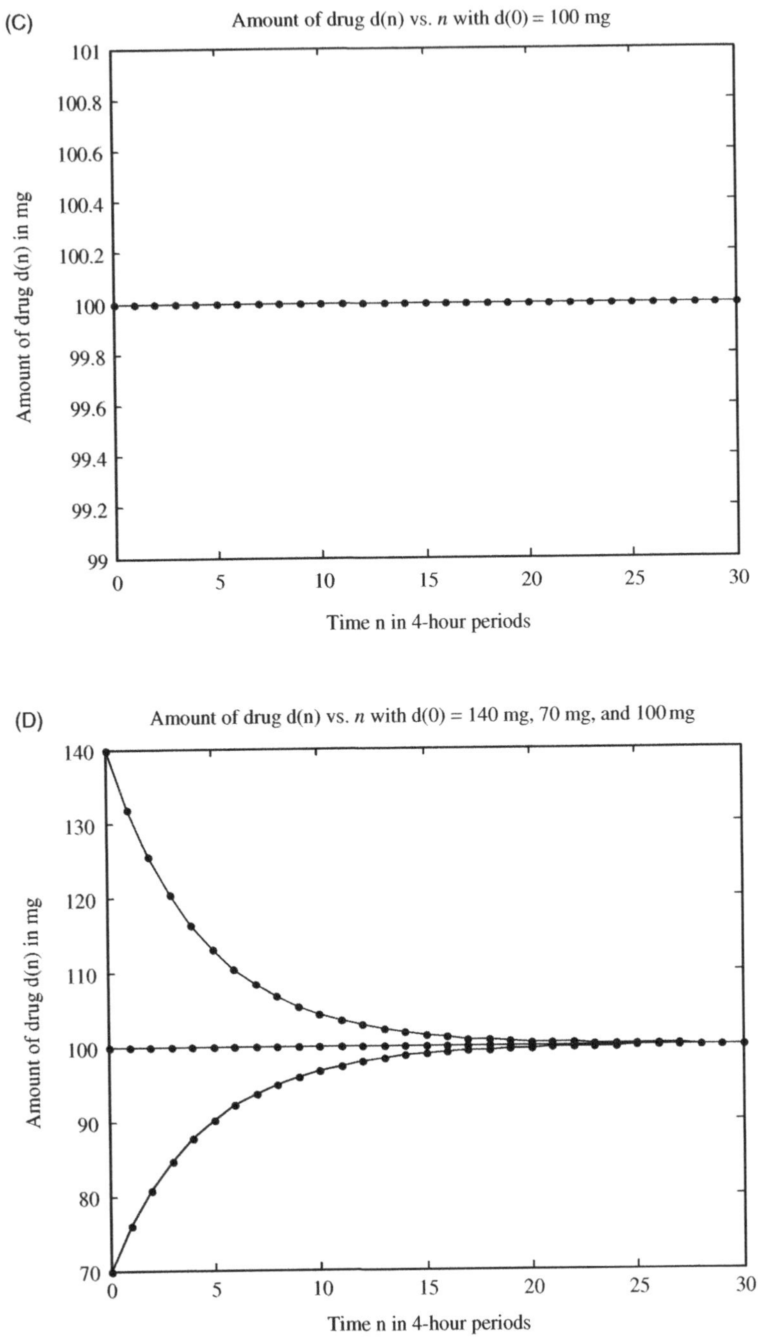

FIGURE 2.3. (*Continued*)

Iterating this system, we get

$$d_1 = 0.8d_0 + 20 = 0.8(100) + 20 = 100.$$
$$d_2 = d_3 = d_4 = d_5 = 100.$$

The amount of the drug after 20 h is 100 mg.

iii. B. The graph of (n, d_n), $n = 0, 1, 2, \ldots, 30$, shown in Figure 2.3C, shows that the amount of drug in the blood is fixed without change at the level of 100 mg. Explain why.

Figure 2.3D shows the three graphs in one coordinate system.

To add more graphs to existing graph\in MATLAB, enter the command **hold on**. The command **hold on** causes all subsequent plot commands to add their plots to the current coordinate system. The command **hold on** remains active until the command **hold off** is entered.

2.2.2. Analytical Solution of a First-Order Linear Difference Equation

We will derive the analytical solution of a first-order linear difference equation with constant coefficients in the form 2.9,

$$y_{n+1} = ay_n + b, \tag{2.9}$$

where a and b are constants.

We have

$$y_1 = ay_0 + b.$$

and

$$y_2 = ay_1 + b = a(ay_0 + b) + b = a^2 y_0 + ab + b.$$

Similarly,

$$y_3 = ay_2 + b = a(a^2 y_0 + ab + b) + b = a^3 y_0 + a^2 b + ab + b.$$

From this pattern we conclude that

$$y_n = a^n y_0 + a^{n-1}b + a^{n-2}b + \ldots + b = a^n y_0 + b(a^{n-1} + a^{n-2} + \ldots + 1).$$
$$y_n = a^n y_0 + b(1 + a + a^2 + \ldots + a^{n-1}) \tag{2.13}$$

Recall that the sum of the first k terms of a geometric sequence $x, rx, r^2x, r^3x, \ldots$ is

$$x + rx + r^2 x + \ldots + r^{k-1} x = \frac{x(r^k - 1)}{r - 1} \quad \text{if } r \neq 1.$$

and if $r=1$, the sum equals kx. Consequently, if $a \neq 1$, equation 2.13 becomes

$$y_n = a^n y_0 + \frac{b(a^n - 1)}{a-1} \tag{2.14}$$

If $a = 1$, equation 2.13 yields

$$y_n = 1^n y_0 + b(1+1+\ldots+1).$$
$$y_n = y_0 + bn. \tag{2.15}$$

Analytically we can check that equation 2.14 is a solution of equation 2.9 by showing that y_n in equation 2.14 satisfies equation 2.9. We have

$$\text{LHS of } (2) = y_{n+1} = a^{n+1} y_0 + \frac{b(a^{n+1} - 1)}{a-1}.$$

$$\text{RHS of } (2) = ay_n + b = a\left(a^n y_0 + \frac{b(a^n - 1)}{a-1}\right) + b.$$

$$= a^{n+1} y_0 + \frac{ab(a^n - 1)}{a-1} + b = a^{n+1} y_0 + \frac{ab(a^n - 1) + b(a-1)}{a-1}.$$

$$= a^{n+1} y_0 + \frac{b(a^{n+1} - 1)}{a-1}.$$

Therefore, LHS of equation 2.9 = RHS of equation 2.9.

For example, the analytical solution of the difference equation 2.10 in Model 2.5, given in situation i, is determined by,

$$y_n = 140(0.8)^n + \frac{20(0.8^n - 1)}{0.8 - 1}.$$
$$y_n = 140(0.8)^n - 100(0.8^n - 1).$$

This analytical solution can be used to determine amount of the drug in the blood after any value for n without iterating the difference equation. For example, to determine the amount of the drug in the blood after 2 days in the above situation, we put $n = 12$ in the equation:

$$y_{12} = 140(0.8)^{12} - 100(0.8^{12} - 1) = 102.75\,\text{mg}.$$

2.2.3. Constant Solutions and Equilibrium Values

In Model 2.5, we realized that if $y_0 = 100$, then $y_1 = 100$, $y_2 = 100, \ldots$. In other words

$$y_n = 100 \quad \text{for } n = 0, 1, 2, \ldots,$$

which means that the values of the solution do not change with time and remain constant at 100. This situation is called a **constant** or **steady-state** solution. The constant value 100 is called an **equilibrium value** of the difference equation. If a solution of a difference equation reaches the equilibrium value over a period of time, it remain constant at the equilibrium value as in Model 2.5 situations i and ii. Note that a difference equation might have more than one equilibrium value. If, as in Model 2.5 situations i and ii, the solutions of a difference equation with initial values greater than or less than the equilibrium value approach the equilibrium value, then the equilibrium value is called **stable** or an **attractor**. An equilibrium value is called **unstable** or a **repeller** if the solutions of a difference equation with initial values less than or greater than the equilibrium value diverge from the equilibrium value.

Let E be a constant representing an equilibrium value of a difference equation 2.9. To determine E put

$$y_n = E \quad \text{and} \quad y_{n+1} = E.$$

in equation 2.9. We obtain

$$E = aE + b.$$

$$(1-a)E = b.$$

$$E = \frac{b}{1-a}. \tag{2.16}$$

Note that the equilibrium value of a first-order linear difference equation with constant coefficients depends on the values of the coefficients and does not depend on the initial values.

For example, the equilibrium value, E, of difference equation 2.10 in Model 2.5 is

$$E = \frac{20}{1-0.8} = 100 \text{ mg},$$

which is the same value determined graphically for Model 2.5.

Solution of a First-Order Linear Difference Equation with Constant Coefficients

The **analytical solution** of the difference equation

$$y_{n+1} = ay_n + b, \tag{2.9}$$

where a and b are constants, is

$$y_n = a^n y_0 + \frac{b(a^n - 1)}{a - 1}, \quad \text{if } a \neq 1 \tag{2.14}$$

and

$$y_n = y_0 + bn, \quad \text{if } a = 1, \tag{2.15}$$

where y_0 is the initial condition.

The **equilibrium value** E of the difference equation 2.9 is

$$E = \frac{b}{1 - a} \tag{2.16}$$

2.2.4. Model 2.6: Population Dynamics Revisited

Assume that the annual growth rate of deer in Luzern County is 4% and the state restricts hunting to 8,000 deer every year. Let P_0 be the deer population in 2010 and P_n be the deer population after n years from 2010.

i. Model this situation by a difference equation. Find the model's equilibrium value if it exists.
ii. For each of the following initial populations, generate the numerical solution (n, P_n), $n = 0, 1, 2$, graph it and describe the graph: (a) 250,000, (b) 150,000, and (c) 200,000.
iii. Determine whether the equilibrium value is stable or unstable.

Discussion

i. The deer population after $n+1$ years, P_{n+1}, equals the population at the nth year, P_n, plus the natural growth, 4% of P_n and minus 8,000 deer. We obtain

$$P_{n+1} = P_n + 0.04P_n - 8{,}000.$$

$$P_{n+1} = 1.04P_n - 8{,}000. \tag{2.17}$$

This situation is modeled by the above first-order linear nonhomogenous difference equation with constant coefficients.

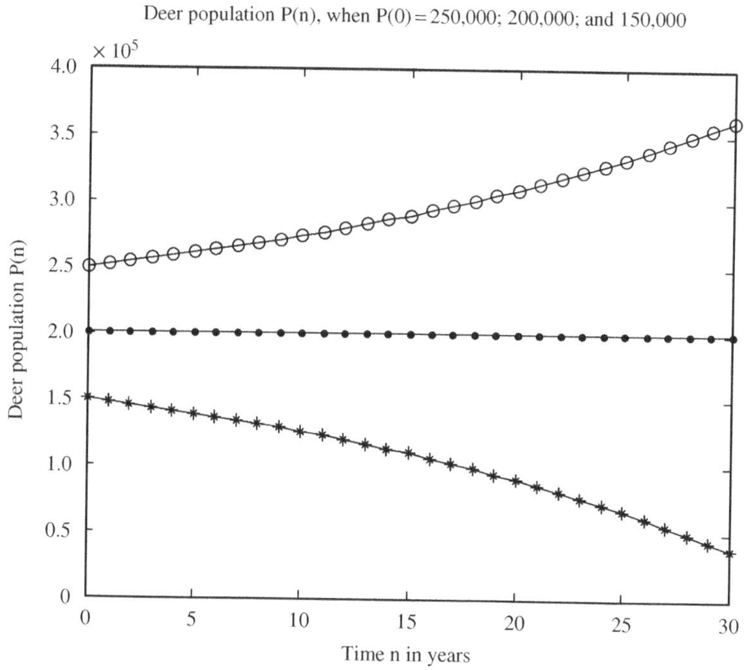

FIGURE 2.4. Three graphs of the deer population, d_n, vs. the time, n, in years, with the initial populations $P_0 = 150{,}000$, $P_0 = 200{,}000$, and $P_0 = 250{,}000$, and $n = 0, 1, \ldots, 30$, for the difference equation $P_{n+1} = 1.04P_n - 8{,}000$.

Let E be a constant representing the equilibrium value of the difference equation 2.17. From equation 2.16 we have

$$E = \frac{b}{1-a} = \frac{-8{,}000}{1-1.04} = 200{,}000.$$

Thus the equilibrium value is 200,000 deer, which means that if the initial population is 200,000 or the population reaches 200,000 the deer population will stay constant at 200,000 without change.

ii. The graphs for the numerical solutions (n, P_n), $n = 0, 1, 2, \ldots, 30$ for the initial populations $P_0 = 250{,}000$, $P_0 = 150{,}000$, and $P_0 = 200{,}000$ are shown in Figure 2.4.

The following MATLAB code will graph the difference equation 2.17 with $P_0 = 250{,}000$. Create a new script file and save the file.

```
t=0;
p=250000;
M=[t p];
```

```
for k=1:30
    p=1.04*p-8000;
    M= [M; k p];
end;
plot(M(: , 1), M(: , 2), 'k.', M(: , 1), M(: , 2), 'k')
xlabel ('Time n in years')
ylabel ('Deer population P(n)')
title('Deer population P(n) with P(0)=250,000;
200,000; and 150,000');
```

Use the command **hold on** and change the initial population (the second line) to **p=150000** and run the modified script file. Do the same thing for P_0=200,000. The three graphs are shown in Figure 2.4.

Another version of this script file can be written:

```
T=0:30;              % Create array T of 31 slots for time
P=zeros(1, 31);      % Create 31-slot array P of zeros for
                     % population
P(1)=250000;         % Store the initial population in the
                     % first slot
                     % Note that p0=250000 is stored in
                     % the P(1)
for k=1:30
    P(k+1)=1.04*P(k)-8000;
end;
plot (T, P, 'ko', T, P, 'k');
label ('Time n in years');
ylabel ('Deer population P(n)');
title('Deer population P(n) with P(0)=250,000;
200,000; and 150,000');
```

1. For P_0=250,000, the numerical solution and its graph show that P_n is increasing without bound. This means that as n increases, P_n increases and when $n \to \infty$, $P_n \to \infty$.
2. The numerical solution for P_0=150,000 and its graph show that the deer population decreases. Eventually, the population will be zero, which means that the deer will be extinct.
3. For P_0=200,000, as you expect the solution is constant at 200,000.

This model is unrealistic. We will study more realistic population dynamics models in the subsequent sections.

iii. The graphs in Figure 2.4 conclude that for any initial condition greater than or less than the equilibrium value, the solution diverges from the equilibrium value. Consequently, the equilibrium value is unstable or a repeller.

2.2.5. Model 2.7: Drugs Revisited

i. Suppose that the kidneys remove 20% of a drug from the blood every 4h. Assume that a patient takes an initial dose of the drug followed by a dose of the same drug every 4h. Determine the additional dose every 4-h period to have the equilibrium value of the drug in the blood be 120mg.

ii. Suppose that the kidneys remove 20% of a drug from the blood every 4h. Assume that a patient takes an initial dose of the drug followed by a dose of 10mg of the same drug after the first 4h and then this dose is increased by 10% of the previous dose every 4h. Model this situation. Explore the behavior of the numerical solutions of the model for different initial doses for the first 5 days. Make a conjecture and explain it.

Discussion

i. Let d_n be the amount of drug in the blood after n 4-h periods and x be the additional dose every 4-h period. This situation is modeled by the difference equation

$$d_{n+1} = 0.8d_n + x.$$

The equilibrium value E of this equation is

$$E = \frac{x}{1-0.8}.$$

Because $E = 120$, we have

$$x = 0.2(120) = 24\,\text{mg}.$$

ii. Let d_n denote the amount of drug in the blood at the end of n 4-h periods, d_0 denote the initial amount of drug in the blood—that is, the initial dose of the drug—and b_n represent the additional dose at the end of the nth 4-h period.

We want to find a relationship between d_{n+1} and d_n. Because at the end of the $(n+1)$th 4-h period, 20% of d_n is removed and the variable dose is added, we have

$$d_1 = d_0 - 0.2d_0 + 10 = 0.8d_0 + 10.$$

$$d_2 = d_1 - 0.2d_1 + 10 + 0.1(10) = 0.8d_1 + 10(1.1).$$

$$d_3 = 0.8d_2 + 10(1.1)^2.$$

Similarly,

$$d_{n+1} = 0.8d_n + 10(1.1)^n, \quad n = 0, 1, 2, \ldots, k. \tag{2.18}$$

Equation 2.18 is a first-order linear difference equation with constant coefficients in the form

$$y_{n+1} = ay_n + b_n.$$

where $a=0.8$ is constant and $b_n=10(1,1)^n$ is variable (depends on n).

In the exercises for this section you will be asked to use MATLAB to graph (n, d_n), $n = 0, 1, 2, \ldots, 24$ (24 4-h intervals in 4 days) with different initial conditions.

2.2.6. Model 2.8: Forensic Application of Newton's Law of Cooling

Let's consider the following situation. A forensic scientist was called to investigate the case of a person found murdered in a room at an office building. She arrived at the scene and started her investigation by measuring the temperature of the body. At 10:00 P.M. the temperature was 80°F and 1 hr later, at 11:00 P.M., she measured the temperature again and it was 76°F. The room temperature where the body was found was set at 70°F. One of the main tasks for the forensic scientist is to determine the time of death.

Discussion

Newton's law of cooling states that the change of the temperature of a cooling object is proportional to the difference between the temperature of the object and the temperature of the object's surroundings. Let

$$\Delta T_n = T_{n+1} - T_n,$$

where T_n is the object's temperature after n time periods and T_0 is the initial temperature. Newton's law is represented by the expression

$$\Delta T_n \propto (R_n - T_n),$$

where R_n is the surrounding temperature. This means that

$$T_{n+1} - T_n = \beta(R_n - T_n), \qquad (2.19)$$

where β is a constant.

In our situation we assume that T_0 is the body's temperature at the murder time. Because the temperature of normal person is 98.6°F, we assume that $T_0=98.6$, and because the room temperature is kept at 70°F, we assume that R_n is a constant with $R_n= 70$. We have

$$T_{n+1} - T_n = \beta(70 - T_n).$$

$$T_{n+1} = (1-\beta)T_n + 70\beta. \qquad (2.20)$$

Equation 2.10 is a first-order linear difference equation with constant coefficients. Equation 2.14 gives the analytical solution of 2.20 in the form

$$T_n = 98.6(1-\beta)^n + \frac{70\beta((1-\beta)^n - 1)}{(1-\beta)-1}.$$

$$T_n = 70 + 28.6(1-\beta)^n. \tag{2.21}$$

Let $T_n = 80$ and $T_{n+1} = 76$ be the body's temperature after n and $n+1$ h from the murder time, respectively. From equation 2.21 we have

$$80 = 70 + 28.6(1-\beta)^n,$$

and

$$76 = 70 + 28.6(1-\beta)^{n+1}.$$

By simplification we have

$$(1-\beta)^n = \frac{10}{28.6} \quad \text{and} \quad (1-\beta)^{n+1} = \frac{6}{28.6}.$$

Thus

$$\frac{(1-\beta)^{n+1}}{(1-\beta)^n} = \frac{6}{10}.$$

$$1-\beta = 0.6.$$

Therefore, $\beta = 0.4$. Substituting β in equation 2.21, we get

$$T_n = 70 + 28.6(1-0.4)^n.$$

$$T_n = 70 + 28.6(0.6)^n.$$

Substituting $T_n = 80$ we get

$$80 = 70 + 28.6(0.6)^n.$$

$$(0.6)^n = \frac{10}{28.6}.$$

Solve for n by taking log of both sides of the this equation, we get

$$\log(0.6)^n = \log\left(\frac{10}{28.6}\right).$$

$$n\log(0.6) = \log\left(\frac{10}{28.6}\right).$$

$$n = \log(10/28.6)/\log(0.6) = 2.05.$$

Therefore, the victim was murdered 2 h and 3 min before 10:00 P.M., which is 7:57 P.M.

2.2.7. Exercises 2.2

1. For the following difference equations figure out by hand y_1 through y_4.

   A. $y_{n+1} = 2y_n - 10,\ y_0 = 12$
   B. $y_{n+1} = 0.5y_n + 10,\ y_0 = 40$
   C. $y_{n+1} = 1.5y_n + 4,\ y_0 = 16$
   D. $y_{n+1} = 0.8y_n - 2,\ y_0 = 20$

   In Exercises 2–5,

   A. Calculate, without computer, the first four terms after the initial condition, y_1, y_2, y_3, and y_4.
   B. Find the analytical solution of the difference equation and use it to find y_4, y_{10}, and y_{20}.

2. $y_{n+1} = 0.8y_n + 100,\ y_0 = 500.$

3. $y_{n+1} = 0.85y_n + 30,\ y_0 = 400.$

4. $y_{n+1} = 1.06y_n - 400,\ y_0 = 10{,}000.$

5. $y_{n+1} = 0.1y_n - 440,\ y_0 = 4{,}000.$

   In Exercises 6–9,

   A. Find the equilibrium value if it exists.
   B. Use MATLAB to find the numerical solution, (n, y_n), $n = 0, 1, 2, \ldots, 20$. Graph the solution.
   C. Determine whether the equilibrium value found in part A is stable or unstable. Recall that one way to determine the stability of an equilibrium value is to investigate the behavior of the solution for initial values greater than and less than the equilibrium value. Explain your answer and sketch the graphs for each equilibrium value on a single coordinate system. Label your graphs.

6. $y_{n+1} = 0.8y_n + 100,\ y_0 = 500.$

7. $y_{n+1} = 0.85y_n + 30,\ y_0 = 400.$

8. $y_{n+1} = 1.06y_n - 400,\ y_0 = 10{,}000.$

9. $y_{n+1} = 0.1y_n - 440,\ y_0 = 4{,}000.$

10. A farmer grows trout in a fish farm. The trout growth rate is 2% per month. Assume that the farmer harvests 2,000 trout at the end of each month. Let P_0 be the initial trout population and P_n be the trout population after n months.

    A. Assuming that the trout population can be modeled by a first-order linear difference equation, model the trout population by a difference equation—that is, find the relationship between P_{n+1} and P_n.

B. Find the analytical solution of the established difference equation in part A and use it to find the trout population after 4 months and after 1 year if the initial population is 112,000 trout.

C. Find an equilibrium value, E, of the difference equation that has a biological meaning.

D. Use MATLAB to describe the long-term behavior of the trout population for different initial populations. Hint: Use MATLAB to find the numerical solution (n, P_n), $n = 0, 1, 2, \ldots, 24$ of the difference equation for $P_0 > E$, and $P_0 < E$. Sketch the graphs.

11. Assume that the laboratory testing on a certain drug used for heart disease patients concluded that every day the body of a patient absorbs 25% of the amount of the drug present in the blood at the beginning of that day. Assume that the patient takes an initial dose of the drug followed by 2 mg of the same drug every day. Let D_n represents the amount of drug in the bloodstream after n days.

A. Model the amount of the drug in the bloodstream by a difference equation.

B. Find the analytical solution of the difference equation.

C. Use the analytical solution to determine the amount of the drug in the bloodstream after 5 days if $D_0 = 12$ mg.

D. Use MATLAB to find a numerical solution of the difference equation with $D_0 = 12$ and determine the amount of drug after 5 days. Be sure that your answers in parts C and D are the same.

E. Find the equilibrium value of the difference equation. Determine whether the equilibrium value is stable or unstable. Explain the meaning of stability/instability in the context of this problem.

12. Consider the administration of drugs discussed in Model 2.7. The model is represented by equation 2.18, $d_{n+1} = 0.8d_n + 10(1.1)^n$, $n = 0, 1, 2, \ldots, k$.

A. Use MATLAB to graph (n, d_n), $n = 0, 1, 2, \ldots, 24$ (24 4-h intervals in 4 days) with different initial conditions such as $d_0 = 30$, 40, 50, 60, 100, 150, and 200.

B. Describe the behavior of the system in the first 4 days.

13. Assume that the environmental scientists have determined that each year 10% of pollutants present in a lake at the beginning of the year is neutralized by solar oxidation. Assume that 10 tons of industrial waste pollutants are dumped into the lake every year. Let y_n be the amount of pollutant (in tons) in the lake after n years from 2010, where y_0 is the amount of pollutant in the lake in 2010.

A. Model the change of the amount of pollutant in the lake by a difference equation—that is find a relationship between y_{n+1} and y_n.

B. Find the analytical solution of the difference equation. Use the analytical solution to predict the amount of pollutants in the lake in 2016 if $y_0 = 80$.

C. Find the equilibrium value E of the difference equation.

D. Use MATLAB to find the numerical solution (n, y_n), $n = 0, 1, 2, \ldots, 20$ and graph it for three initial values:

i. $y_0 > E$

ii. $y_0 < E$

iii. $y_0 = E$

Sketch the three graphs on a single coordinate system. Determine the stability of the equilibrium value and its meaning in the context of this situation.

14. A forensic scientist was called to investigate the case of a person found murdered in a room at an office building. She arrived at the scene and started her investigation by measuring the temperature of the body. At 1:00 A.M. the temperature was 84°F and 1 h later (at 2:00 A.M.), she measured the temperature again, and it was 81°F. The room temperature where the body was found was set at 68°F. Determine the time of death.

15. Police were called to investigate a murder and arrived at the scene to find a murdered employee on a factory floor. At 4:00 A.M. they started their investigation by measuring and recording the body's temperature, which was 89°F. At 5:00 A.M., they recorded a drop in the body's temperature to 87°F. The police investigators noticed also that the temperature at the crime scene was kept constant at 80°F. Determine the time of death.

16. Detective Nancy went to an apartment tower to investigate a murder. To determine the time of death the first thing she did was to measure and record the temperatures of the body and the room where the body was found. At 8:00 A.M. the body temperature was 79°F and the room was 71°F. At 9:00 A.M. Nancy measured and recorded the temperatures of the body and the room again. These were 77°F and 71°F, respectively. In addition she noticed that the thermostat in the room where the body was found had been set at 71°F. After using her computer to make computations, Nancy indicated to her investigative team the time of the victim's death. What was it?

2.3. MODELING WITH NONLINEAR DIFFERENCE EQUATIONS: LOGISTIC GROWTH MODELS

In Sections 2.1 and 2.2, we explored models that were represented by linear difference equations. There are many situations that cannot by modeled by linear difference equations. In this section, we will investigate some models represented by nonlinear difference equations. In particular, we will focus on models represented by logistic difference equations.

2.3.1. Linear Equations

In Section 2.1 (see Model 2.6) we discussed unrestricted population growth of a single species. We assumed that there is no limitation on food and space, and we ignored the interaction of the species with other species. We also considered the species in a closed environment that has no immigration or emigration. Let's review the population model of a single species represented by a single first-order linear difference equation.

Let P_n be the population of a certain species at the end of the nth time period (e.g., breeding season) and let a be the growth rate per period. We have

$$P_{n+1} = P_n + aP_n.$$

$$P_n = (1+a)P_{n-1}. \tag{2.22}$$

Letting $r = 1 + a$, we get

$$P_n = rP_{n-1}, \tag{2.23}$$

which is a first-order linear difference equation with constant coefficients. The number of individuals in the next generation per individual in the current generation, r, is called the **growth parameter** (or **factor**). The solution of the difference equation 2.23 is

$$P_n = r^n P_0, \tag{2.24}$$

where P_0 is the initial population, $P_0 > 0$. This solution expresses P_n as an exponential function of n. If $a>0$, $r>1$, and for large values of n, P_n increases exponentially without bound. In other words, the population explodes, which is unrealistic. If $a < 0$, P_n approaches 0—that is, the population dies out. On the other hand, if $a = 0$, there is no change in the population, and $P_n = P_0$ for every n. In general, the environment and the food supply do not support an unbounded increase in the population; therefore, there is a need for a more realistic model that takes into consideration the fact that the available resources can support only a limited population.

2.3.2. Logistic Equations

It was noted that a linear model for population growth, which predicts exponential growth of the population, fits a small population. However, the population tends to level off to a limiting value over a period of time due to a limitation in the available resources. Therefore, the exponential growth model represented by a linear difference equation needs to be modified to incorporate the new factors. This suggests a model that would predict an exponential growth for the first k populations $P_0, P_1, \ldots, P_k$, where the populations are small, and then level off to a limiting value. In 1838,

Pierre-Francois Verhulst (1804–1849), a Dutch mathematical biologist, suggested replacing equation 2.22 with

$$P_{n+1} = P_n + aP_n - bP_n^2, \tag{2.25}$$

where b is a very small positive constant compared to the constant a. This model suggests subtracting a positive term from the right-hand side of equation 2.22. Equation 2.25 is a first-order difference equation. However, this equation is nonlinear because P_n is raised to the power 2. Equation 2.25 is called a **logistic equation** and a population model represented by it is called a **logistic growth model**. In general, there is no analytical solution of the nonlinear difference equation 2.25. Therefore, we will use MATLAB to find a numerical solution of a given nonlinear difference equation.

Let's first explore the logistic equation 2.25 for specific values of a, b, and P_0.

2.3.3. Model 2.9: Logistic Population Dynamics

Consider the logistic equation 2.25, with $a=0.4$ and $b=0.0002$. For each of the following initial populations, find the numerical solution (n, P_n), $n = 0, 1, 2, \ldots, 30$, graph the solution, and describe the graph:

i. $P_0 = 200$
ii. $P_0 = 2{,}800$
iii. $P_0 = 2{,}000$

Discussion

i. The difference equation 2.25 becomes

$$P_{n+1} = 1.4P_n - 0.0002P_n^2 \tag{2.26}$$

You may use one of the following MATLAB codes to create the numerical solution of this nonlinear difference equation and graph it.

Version 1

```
n=0;
p=200;
M=[n p];
for k=1:30
    p=1.4*p-0.0002*p^2;
    M=[M; k p];
end;
plot(M(: , 1), M(: , 2), 'k.');
xlabel ('Time n');
ylabel ('Species population P(n)');
```

Version 2

```
n=0:30;
P=zeros(1, 31);
P(1)=200;
for k=1:30
    P(k+1)=1.4*P(k)-0.0002*P(k)^2;
end;
plot(n, P, 'k.', n, P, 'k');
xlabel ('Time n');
ylabel ('Species population P(n)');
```

The graph is shown in Figure 2.5A. The graph shows that the population increases and then levels off to 2000 and remains at that level. In particular, for the first 13 time periods, the population grows rapidly in an exponential fashion, then grows slowly and levels off to the value 2000 and remains at that value.

ii. Iterate equations 2.26 with the initial value $P_0=2{,}800$ to generate numerical solution and graph it. The graph of the numerical solution is shown in Figure 2.5B. The graph shows that the population rapidly decreases to the limiting value 2000. In this situation, the initial population is greater than the limiting value, which means that the available resources do not support this population. This leads to a decrease in the population until it reaches the limiting value and stays there without change.

iii. Iterating the difference equation 2.26 with the initial value $P_0=2{,}000$ produces a constant solution,

$$P_n = 2000, \quad \text{for } n = 0, 1, 2, \ldots$$

The graph is shown in Figure 2.5C. In this situation, where the initial population $P_0=$ the limiting value 2000 all populations stay without change at the limiting value of 2000.

The three graphs in a single coordinate system are shown in Figure 2.5D. This can be accomplished by adding the command **hold on** and graphing each situation.

2.3.4. Carrying Capacity

The value 2000 in Model 2.9 is the population that the given environment and the available resources can support. This value is called the **carrying capacity** or the **limiting value**.

Let's see how to determine the limiting values of the logistic difference equation 2.25.

$$P_{n+1} = P_n + aP_n - bP_n^2. \tag{2.25}$$

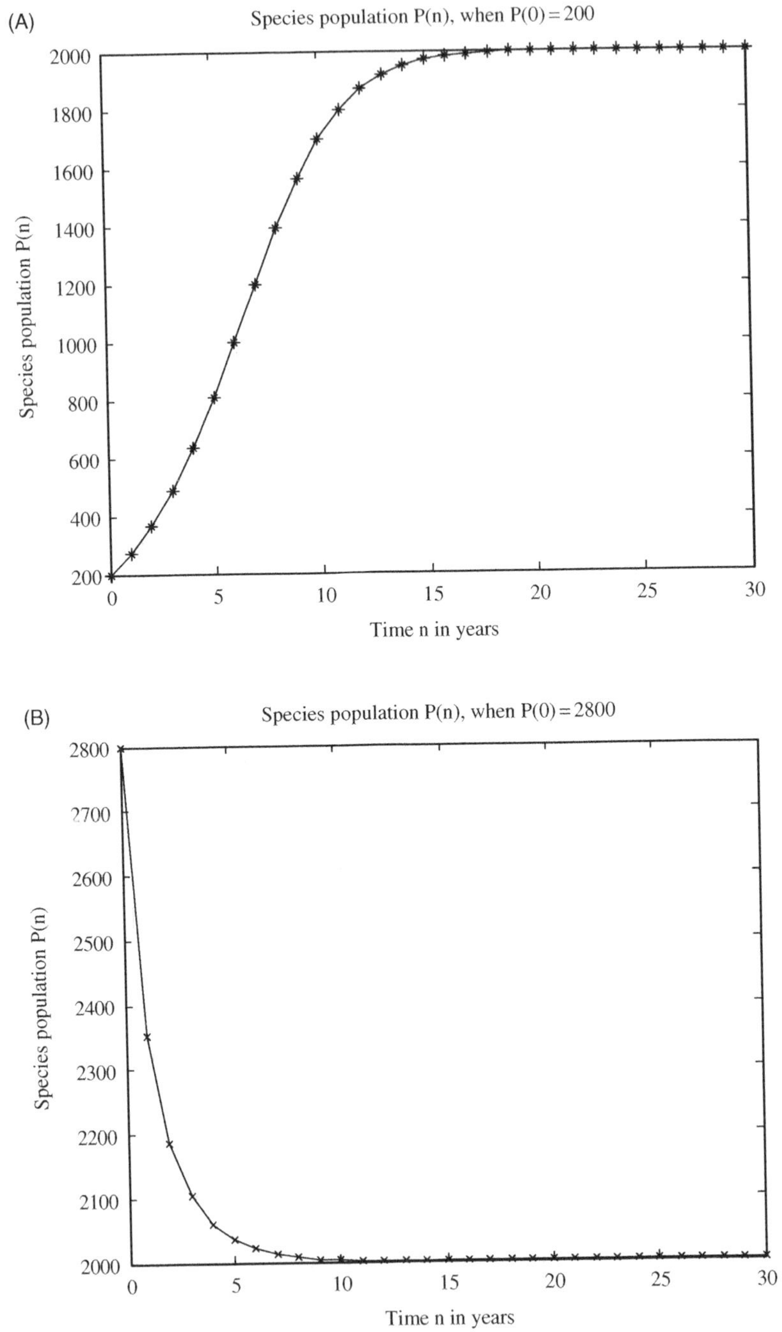

FIGURE 2.5. Graphs of a species population, P_n, vs. time, n, in years for the difference equation $P_{n+1} = 1.4P_n - 0.0002P_n^2$. **A.** $P_0 = 200$ and $n = 0, 1, 2, \ldots, 30$. **B.** $P_0 = 2{,}800$ and $n = 0, 1, 2, \ldots, 30$. **C.** $P_0 = 2{,}000$ and $n = 0, 1, 2, \ldots, 30$. **D.** $P_0 = 200$, $P_0 = 2{,}000$, and $P_0 = 2{,}800$ for $n = 0, 1, 2, \ldots, 30$.

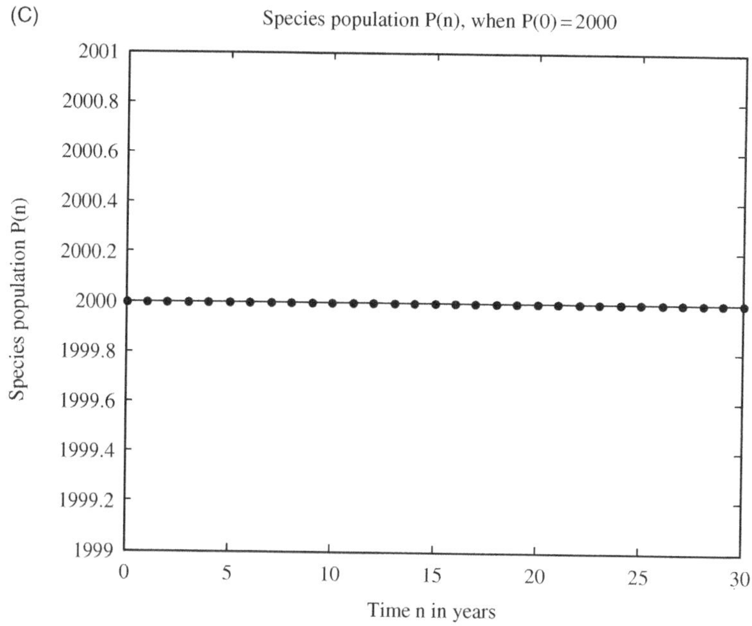

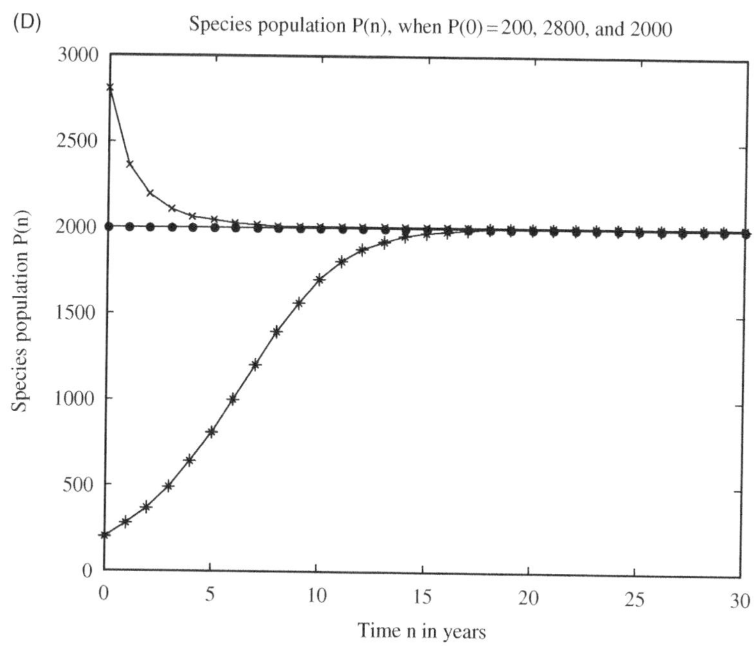

FIGURE 2.5. (*Continued*)

Let L be a limiting value. At a limiting value L, we have $P_n = P_{n+1} = L$. Consequently, to determine the limiting values, put $P_n = L$ and $P_{n+1} = L$ in the difference equation and solve for L. By substitution in equation 2.25, we get

$$L = L + aL - bL^2.$$
$$L(a - bL) = 0,$$

which implies that

$$L = 0 \quad \text{or} \quad L = \frac{a}{b}. \tag{2.27}$$

Note that $L = 0$ is equivalent to $P_n = 0$, which means that there is no population.

As an example for determining the limiting value, L, for the difference equation in Model 2.9,

$$P_{n+1} = P_n + 0.4P_n - 0.0002P_n^2,$$

we have

$$L = \frac{a}{b} = \frac{0.4}{0.0002} = 2{,}000.$$

Note that this value is the same as the value determined graphically.

Let's write the logistic equation 2.25 in the form

$$P_{n+1} - P_n = aP_n - bP_n^2.$$

The left-hand side of this equation, $P_{n+1} - P_n$, represents the change in the population over one time period. For the purpose of investigating the change in the population, we will factor the right-hand side of the equation. We have

$$P_{n+1} - P_n = aP_n - bP_n^2.$$
$$P_{n+1} - P_n = aP_n\left(1 - \frac{bP_n}{a}\right).$$
$$P_{n+1} - P_n = aP_n\left(1 - \frac{P_n}{L}\right), \tag{2.28}$$

where $L = \frac{a}{b}$. We will consider equation 2.28 in three situations.

Situation 1: $P_n < L$

In this case, $\frac{P_n}{L} < 1$ and the right-hand side of the equation 2.28 is positive. Consequently, $P_{n+1} - P_n$ is positive—that is $P_{n+1} > P_n$. In other words, P_n increases and approaches L. As P_n approaches L, $\left(1 - \frac{P_n}{L}\right)$ becomes very small and approaches zero. Consequently, the change in population $P_{n+1} - P_n$ approaches zero, which means that $P_{n+1} = P_n$ for large n, as shown in Figure 2.5A.

Situation 2: $P_n > L$

We have $\frac{P_n}{L} > 1$. Thus the right-hand side of the equation 2.28 is negative. Consequently, $P_{n+1} - P_n$ is negative, or $P_{n+1} < P_n$. This means that the population decreases and approaches L. This situation is shown in Figure 2.5B.

Situation 3: $P_n = L$

Because $\frac{P_n}{L} = 1$, the right-hand side of equation 2.28 is zero and the equation becomes,

$$P_{n+1} - P_n = 0.$$

This means that there is no change in the population, and the population is fixed at L, as given in Figure 2.5C.

2.3.5. A Model of Logistic Population Growth with Harvesting

Consider a population of a species that is harvested, such as fish, or hunted, such as deer. We will study one harvest or hunting strategy and evaluate it. We are interested in optimizing the economic yield and in maintaining it. The following model is applicable to any renewable natural resource that is harvested, and the population growth is described by a difference equation. Difference equations have been used efficiently in fishery management.

Maximum Sustainable Yield

Let P_n be the population of the species after n time periods. We assume that the population growth is modeled by logistic difference equation 2.28, where a is the unrestricted growth rate and L is the carrying capacity (limiting value) of the environment. The left-hand side of equation 2.28, $P_{n+1} - P_n$, represents the change in the population. If $P_{n+1} > P_n$, $P_{n+1} - P_n > 0$, and we have an increase in the population. We wish to harvest this increase and maintain the population so the species can replenish itself. The largest population increase is called the **maximum sustainable yield**.

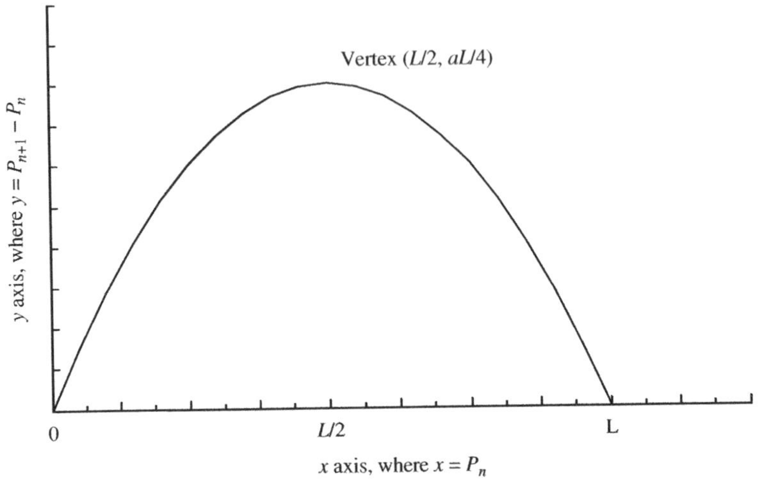

FIGURE 2.6. Graph of the quadratic function $y=ax(1-x/L)$, where $y=P_{n+1}-P_n$, $x=P_n$. The vertex of the parabola is $(L/2, aL/4)$.

The natural questions that arise are

1. At which population does the maximum sustainable yield occur?
2. What is the maximum sustainable yield?

To answer these questions let's consider equation 2.28, Because a and L are constants, the change in the population $P_{n+1}-P_n$ depends on P_n. Let's replace $P_{n+1}-P_n$ with a variable, call it y, and replace P_n with a variable, call it x. We have

$$y = ax\left(1-\frac{x}{L}\right).$$

It is clear that y is a quadratic function of x. Figure 2.6 shows the graph of this function, which is a parabola opened downward. The graph intersects with x-axis at $x = 0$ and $1-\frac{x}{L}=0$ that is $x = L$. Because the vertex of this parabola is the highest point on the curve, we are interested in the coordinates of the vertex. Let (x_v, y_v) be the vertex of the parabola. The x coordinate of the vertex x_v is halfway between the two x intercepts. Therefore

$$x_v = \frac{L}{2}.$$

To find the y coordinate of the vertex put $x = \frac{L}{2}$ in the this equation. We get

$$y_v = a\frac{L}{2}\left(1-\frac{L/2}{L}\right) = \frac{aL}{4}.$$

Thus the y coordinate of the vertex is

$$y_v = \frac{aL}{4}.$$

The maximum sustainable yield, call it Y_m, equals to y_v. Thus

$$Y_m = \frac{aL}{4} \tag{2.29}$$

The maximum sustainable yield occurs at x_v—that is, at population P_n,

$$P_n = \frac{L}{2} \tag{2.30}$$

Example 2.1

Consider the deer population in a closed region with limited resources and assume that the population is modeled by a logistic difference equation. Assume that the annual natural unrestricted growth rate is 10% and the carrying capacity is 100,000. What is the maximum sustainable yield Y_m and at which population P_n the maximum sustainable yield occurs?

Discussion

The deer population in this situation is modeled by the logistic difference equation 2.28, with $a=0.10$ and $L=100{,}000$. We have

$$P_{n+1} - P_n = 0.1P_n\left(1 - \frac{P_n}{100{,}000}\right).$$

The maximum sustainable yield Y_m is

$$Y_m = \frac{aL}{4} = \frac{0.1(100{,}000)}{4} = 2{,}500 \text{ deer.}$$

The population that gives the maximum sustainable yield P_n is

$$P_n = \frac{L}{2} = \frac{100{,}000}{2} = 50{,}000 \text{ deer.}$$

In other words, when the deer population is 50,000, a maximum sustainable yield of 2,500 deer is obtained.

2.3.6. Fixed Harvest Strategy

Assume that the population growth of a species is modeled by the logistic difference equation 2.25. Assume that a fixed number, call it H, of the species is harvested or hunted every time period. We assume that the harvest occurs at the end of each time period in order to not affect the growth rate of the species. Equation 2.25 must be modified to equation 2.31,

$$P_{n+1} = P_n + aP_n - bP_n^2 - H. \tag{2.31}$$

We will investigate the situation in three cases:

Case 1. $H < Y_m$
Case 2. $H > Y_m$
Case 3. $H = Y_m$

Case 1 is investigated in Model 2.10, where we assume that the fixed number of deer harvested every year is smaller than the maximum sustainable yield. In the exercises for this section you will investigate cases 2 and 3.

Model 2.10: Population with Fixed Harvest Dynamics

Consider the deer population modeled by the logistic difference equation in Example 2.1, where the growth rate is 10% and the carrying capacity is 100,000. Recall that the maximum sustainable yield, Y_m, has been determined by 2,500 deer. Assume that 1,600 deer are allowed to be hunted every year.

i. Model this situation by a difference equation.
ii. Find the equilibrium values (fixed points) of the difference equation.
iii. For each equilibrium value that has a biological meaning, determine whether it is stable, unstable, or semistable. Interpret the stability of each equilibrium value.

Discussion

i. Recall that the deer population without hunting was modeled by the logistic difference equation 2.28, with $a = 0.1$ and $L = 100{,}000$:

$$P_{n+1} - P_n = 0.1P_n\left(1 - \frac{P_n}{100{,}000}\right).$$

Now with 1,600 deer being hunted each year, the model must be modified to

$$P_{n+1} - P_n = 0.1P_n\left(1 - \frac{P_n}{100{,}000}\right) - 1{,}600.$$

$$P_{n+1} = 1.1P_n - 0.000001P_n^2 - 1{,}600 \tag{2.32}$$

Thus, this situation is modeled by the first-order nonlinear difference equation (2.32).

ii. To find the equilibrium values, call it E, of equation 2.32, put $P_n = E$ and $P_{n+1} = E$ in the equation and solve for E. We have

$$E = 1.1E - 0.000001E^2 - 1{,}600.$$
$$0.000001E^2 - 0.1E + 1{,}600 = 0.$$

To solve this polynomial we use the following MATLAB command

```
>>roots([0.000001 -0.1 1600])
1.0e+4*
  8.0000
  2.0000
```

Therefore, this quadratic equation has two real roots: $E_1 = 80{,}000$ and $E_2 = 20{,}000$.

Thus the difference equation 2.32 has two equilibrium values $E_1 = 80{,}000$ and $E_2 = 20{,}000$. Figure 2.7A shows the graph of the numerical solution (n, P_n), $n = 0, 1, 2, \ldots, 50$ of equation 2.32, with the initial values $P_0 = 80{,}000$ and $P_0 = 20{,}000$.

iii. Recall that an equilibrium value is stable if the solutions of the difference equation, with initial values are greater than or less than the equilibrium value, converge to the equilibrium value. Thus for the equilibrium value 80,000, we will create the numerical solution (n, P_n), $n = 0, 1, 2, \ldots, 50$ of equation 2.32 for some specific initial values and graph it. Five graphs are shown in Figure 2.7B. The graphs from the lowest to the highest are with the initial values 75,000; 79,000; 80,000; 82,000; and 85,000, respectively. Because these solutions converge to 80,000, the equilibrium value $E_1 = 80{,}000$ is stable. To determine the stability of the equilibrium value $E_2 = 20{,}000$, we created the numerical solution (n, P_n), $n = 0, 1, 2, \ldots, 50$, with some initial values close to 20,000 and graphed it. These graphs are shown in Figure 2.7C. The graphs from the lowest to the highest are with the initial values 19,000; 20,000; and 21,000, respectively. Because the graphs for initial values close to the equilibrium value 20,000 diverge from it, the equilibrium value $E_2 = 20{,}000$ is unstable. Figure 2.7D combines the graphs in panels B and C.

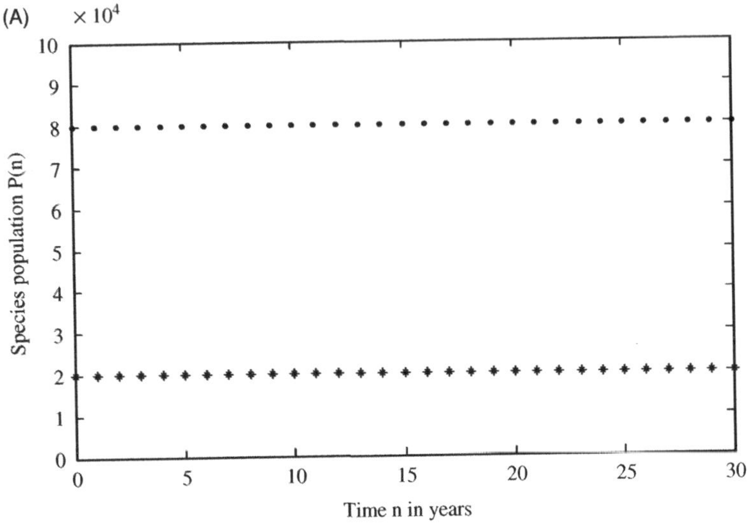

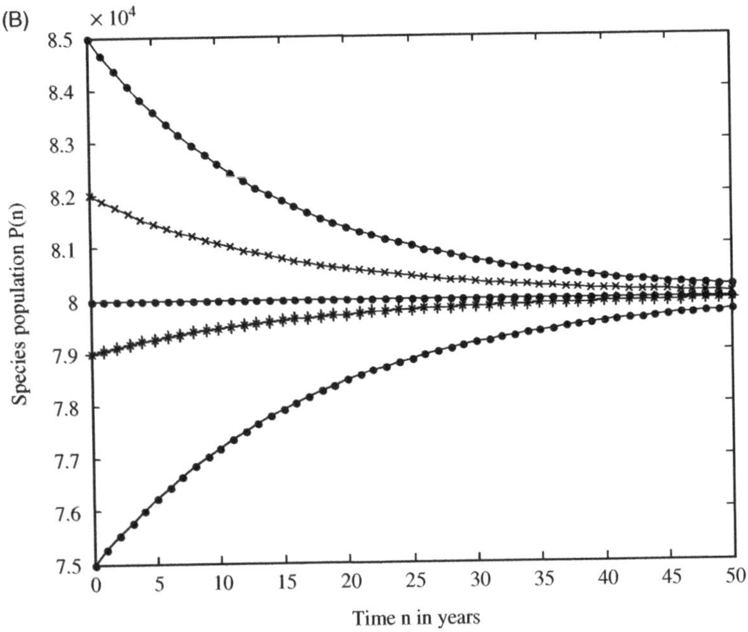

FIGURE 2.7. Graphs of the numerical solution (n, P_n), $n=0,1,\ldots,50$ of the difference equation $P_{n+1}=1.1P_n-0.000001P_n^2-1{,}600$. **A**. Initial values $P_0=20{,}000$ and $P_0=80{,}000$. **B**. Initial values $P_0=75{,}000$, $P_0=79{,}000$, $P_0=80{,}000$, $P_0=82{,}000$, and $P_0=85{,}000$. **C**. Initial values $P_0=19{,}000$, $P_0=20{,}000$, and $P_0=21{,}000$. **D**. A combination of the graphs in panels B and C in one coordinate system.

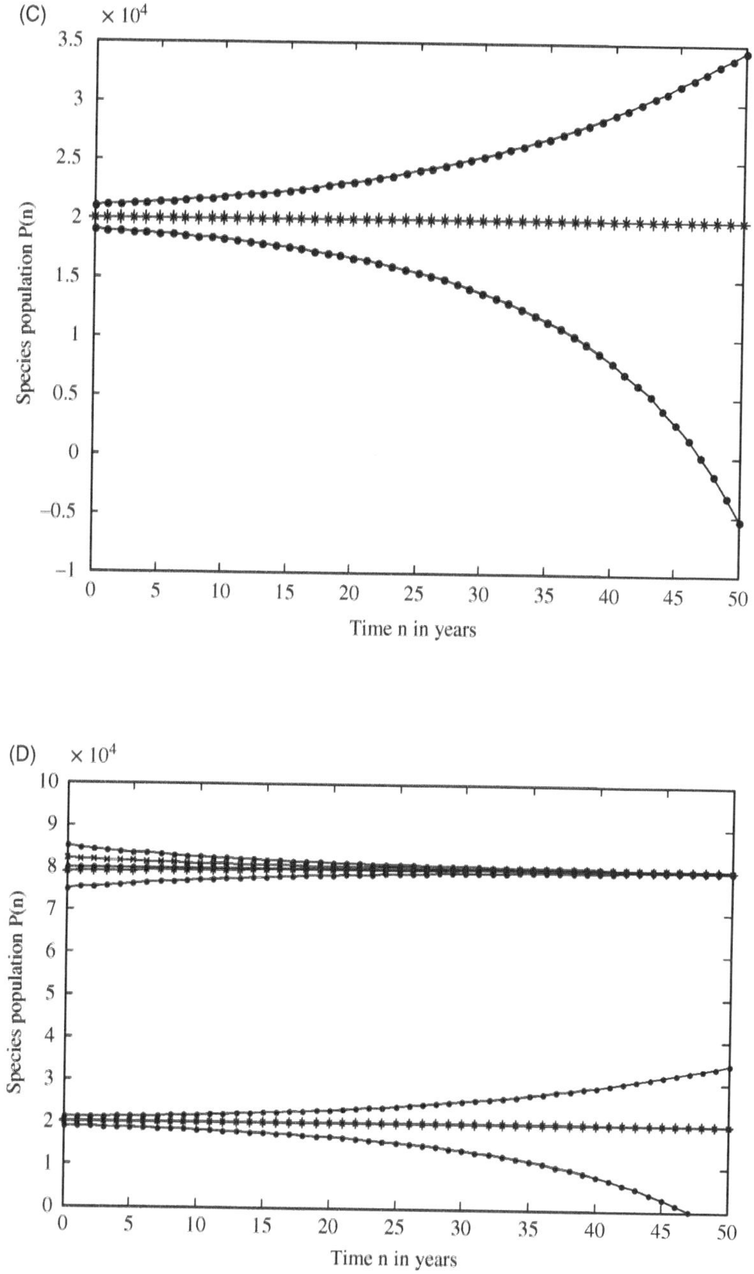

FIGURE 2.7. (*Continued*)

We will interpret the stability of $E_1=80{,}000$ and instability of $E_2=20{,}000$ in four situations:

Situation 1: 20,000 < P_0 < 80,000
In the first situation, the population of deer will increase and converge to the stable equilibrium value $E_1=80{,}000$ and diverges from the unstable equilibrium value $E_2=20{,}000$. Eventually, the population of deer will approach 80,000 and will stay constant at that level. This means that the available resources allow for a natural growth and the natural increase in the population of deer is greater than the number of hunted deer.

Situation 2: 80,000 < P_0 < 100,000
The carrying capacity of the environment in the second situation is 100,000. Because the initial value of deer is close to 100,000, there is a natural increase in deer population. However, this natural increase is less than the number of hunted deer of 1,600. Consequently, the population of deer decreases and converges to the equilibrium value $E_1=80{,}000$.

Situation 3: P_0 > 100,000
Because the initial population is greater than the carrying capacity of the environment in the third situation, there is a natural decline in the population of deer. This decline is combined with the 1,600 deer that are killed every year, resulting in a rapid decrease in the deer population until it levels off at 80,000 and stays constant at that level.

Situation 4: P_0 < 20,000
The natural increase in the deer population in the final situation is less than the number of deer being killed. Consequently, the deer population decreases and reaches 0, and the deer would die out.

The following MATLAB code is used to create the numerical solution of the nonlinear difference equation 2.32 with a specific initial value and graph it.

```
n=0;
p=85000;
M=[n p];
for k=1:50
    p=1.1*p-0.000001*p^2-1600;
    M=[M; k p];
end;
plot(M(:,1), M(:,2), 'k.', M(:,1), M(:,2), 'k')
                                   % To plot the graph in discrete black dots and
                                   % the line graph (connected dots)
```

```
xlabel ('Time n in years')
ylabel ('Species population P(n)')
hold on                     % A new graph will be added to the current graph
```

2.3.7. Exercises 2.3

1. Consider the nonlinear logistic difference equation $P_{n+1} = P_n + aP_n - bP_n^2$, with $a=0.2$ and $b=0.0002$.

   A. What is the carrying capacity of the environment L?

   B. Use MATLAB to find numerical solution (n, P_n), $n = 0, 1, 2, \ldots, 30$ and plot it for the following initial values: $P_0=120$, $P_0=1{,}600$, and $P_0=1{,}000$. Sketch the three plots in a single graph.

2. Repeat problem 1, with $a = 0.3$ and $b = 0.0002$. Select the following three initial populations: $P_0<L$, $P_0>L$, and $P_0 = L$.

3. Repeat problem 1, with $a = 0.2$ and $b = 0.00016$. Select the following three initial populations: $P_0 < L$, $P_0 > L$, and $P_0 = L$.

4. Consider a species modeled by the logistic difference equation $P_{n+1} = P_n + 0.15P_n - 0.0003P_n^2$.

   A. Find the carrying capacity, L, of the environment.

   B. Use MATLAB to find numerical solution (n, P_n), $n = 0, 1, 2, \ldots, 30$ for following three initial values: (i) $P_0<L$, (ii) $P_0>L$, and (iii) $P_0=L$. Sketch the three graphs in a single coordinate system.

5. Repeat problem 4 for the difference equation $P_{n+1} = P_n + 0.02P_n - 0.00004P_n^2$.

6. Consider the nonlinear logistic difference equation $P_{n+1} = P_n + aP_n - bP_n^2$, with $a=0.03$ and the carrying capacity $L=200$.

   A. Find the constant b.

   B. Use MATLAB to find the numerical solution (n, P_n), $n=0, 1, 2, \ldots, 20$ and graph it for the following initial conditions: (i) $P_0=0.25L$, (ii) $P_0=1.4L$, and (iii) $P_0=L$.

7. Repeat problem 6 with $a=0.1$ and $L=400$.

In Exercises 8 and 9, assume that the species population is modeled by the logistic difference equation 2.28.

8. Assume that the marine biologists estimate that the carrying capacity of a whale population in a certain region is 240,000 and the annual unrestricted growth rate is 6%. What is the maximum sustainable yield, Y_m, and at which population, P_n, will the maximum sustainable yield occur?

9. Assume that the trout unrestricted growth rate is estimated by 3% per month and the trout carrying capacity is 100,000. Find the maximum sustainable yield, Y_m, and determine the population that gives the maximum sustainable yield.

In Exercises 10 and 11, consider Model 2.10 with the same parameters, but the value of H is given.

A. Find the equilibrium values of this system.

B. For each equilibrium value that has a biological meaning use MATLAB to determine whether it is stable, unstable, or semistable.

C. Interpret the stability of the equilibrium values found in part B in the context of this situation.

10. $H > Y_m$ (e.g. $H = 2{,}600$).

11. $H = Y_m$ (e.g. $H = 2{,}500$).

In Exercises 12–14, assume that the population growth of a species is modeled by the logistic difference equation 2.25. Assume that a fixed number, call it H, of the species is harvested at the end of each time period. Consequently, the population growth with harvest is modeled by the difference equation 2.31.

12. Consider a whale population in a certain region modeled by the logistic equation 2.25, with $a = 0.06$ and the carrying capacity $L = 240{,}000$. The maximum sustainable yield is $Y_m = 3{,}600$. Now consider the whale population with $H < Y_m$, say $H = 3{,}000$.

A. Find the equilibrium values of this system.

B. For each equilibrium value that has a biological meaning use MATLAB to determine whether it is stable, unstable, or semistable.

C. Interpret the stability of the equilibrium values found in part B in the context of this situation.

13. Repeat problem 12 with $H = Y_m$, say $H = 3{,}600$.

14. Repeat problem 12 with $H > Y_m$, say $H = 3{,}800$.

2.4. LOGISTIC EQUATIONS AND CHAOS

In the last section we studied the logistic equation,

$$P_{n+1} = P_n + aP_n - bP_n^2, \tag{2.25}$$

which is equivalent to

$$P_{n+1} = P_n + aP_n\left(1 - \frac{P_n}{L}\right), \tag{2.28}$$

where P_n is the population of a species after n time periods, a is the unrestricted growth rate per period, b is very small positive constant compared to a, and $L = \frac{a}{b}$ is the carrying capacity of the environment (or the limiting value). We studied equation 2.28 for small values of a, that is $0<a<1$ and we know the behavior of the numerical solution (n, P_n), $n = 0, 1, \ldots, k$ for different initial values of P_0, where $P_0<L$ or $P_0>L$.

In the early 1970s, biologist Robert May discovered that logistic equation 2.28 exhibits fascinating behavior when the growth constant $a > 2$.

Let's rewrite equation 2.25 in the form,

$$P_{n+1} = (1+a)P_n - bP_n^2.$$

$$P_{n+1} = (1+a)P_n\left(1-\frac{b}{1+a}P_n\right). \tag{2.33}$$

For simplicity, we will scale equation 2.33 by the substitution

$$P_n = \frac{1+a}{b}x_n.$$

We have

$$x_{n+1} = (1+a)x_n(1-x_n). \tag{2.34}$$

Letting $\lambda = 1+a$, equation 2.34 becomes

$$x_{n+1} = \lambda x_n(1-x_n). \tag{2.35}$$

If we graph x_{n+1} versus x_n, equation 2.35 is represented by a parabola that opens downward. The x_n intersections are $x_n=0$ and $x_n=1$. The x_n coordinate of the vertex is $\frac{1}{2}$ and the x_{n+1} coordinate of the vertex is $\frac{\lambda}{4}$. Consequently, the coordinates of the vertex are $\left(\frac{1}{2}, \frac{\lambda}{4}\right)$.

As mentioned before, generally, there is no analytical solution of a nonlinear difference equation. Therefore we will use MATLAB to iterate the given difference equation with an initial condition to create a numerical solution. Then we will graph the solution and describe the long-term behavior of the solution. We will see that the solutions of equation 2.35 with $0 < a \le 3$ or equivalently $1<\lambda \le 4$ fall in one of the three categories.

2.4.1. Stable Growth: $0< a \le 2$ or $1 < \lambda \le 3$

Case A. $0<a<1$ or $1<\lambda<2$

Let $a = 0.4$ (i.e., $\lambda = 1.4$) and $x_0=0.2$. Iterate to find (n, x_n), $n = 0, 1, \ldots, 30$ and graph it.

You may use one the following MATLAB codes to produce the numerical solution of the difference equation 2.35 with $\lambda=1.4$ and graph it.

Version 1

```
n=0:30;
x=zeros(1, 31);
x(1)=0.2;
for k=1:30
    x(k+1)=1.4*x(k)*(1-x(k));
end;
plot(n, x, 'k.', n, x, 'k');
xlabel ('Time n');
ylabel ('x(n)');
```

Version 2

```
n=0;
x=0.2;
M=[n x];
for k=1:30
    x=1.4*x*(1-x);
    M=[M; k x];
end;
plot(M(: , 1), M(: , 2), 'k.', M(: , 1), M(: , 2), 'k');
xlabel ('Time n');
ylabel ('x(n)');
```

The graph of the numerical solution is shown in Figure 2.8. The x_n (scaled population) grow in a similar fashion to that of Model 2.9. That is, x_n grows almost exponentially for the first 7 or 8 periods, and then it levels out. Then x_n approaches the limiting value (scaled carrying capacity) of approximately 0.2857 and stays constant at that level.

Case B. $1<a\leq 2$ or $2<\lambda\leq 3$

Let $a=1.8$ (i.e., $\lambda=2.8$) and $x_0=0.2$. Iterate to find (n, x_n), $n=0, 1, \ldots, 30$ and graph it. The graph is shown in Figure 2.9. In the first few time periods x_n grows rapidly and even overshoot the limiting value, then it decreases to a value less than the limiting value. Next x_n increases to a level greater than the limiting value. This behavior continues where x_n oscillates around the limiting value; however, the amplitude of the oscillation decreases. In other words, for $1<a\leq 2$, the solution oscillates around the limiting value and converges to it. This type of oscillation is called **damped oscillation**. The growth of x_n in Cases A and B can be characterized as **stable growth**.

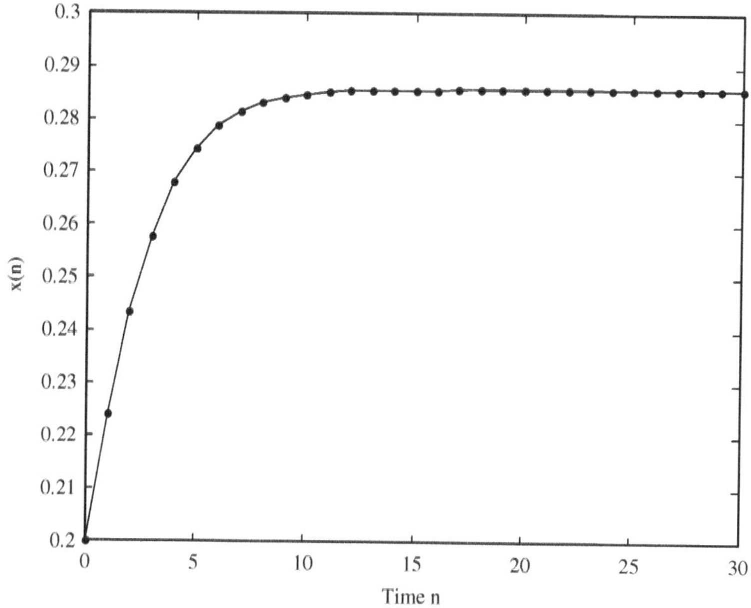

FIGURE 2.8. Graph of the numerical solution (n, x_n), $n = 0, 1, \ldots, 30$ of the difference equation $x_{n+1} = 1.4x_n(1 - x_n)$ with $x_0 = 0.2$.

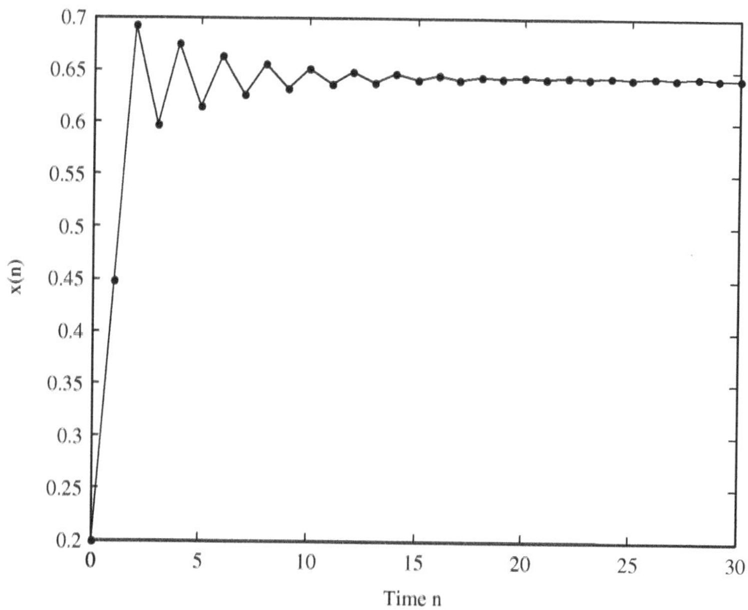

FIGURE 2.9. Graph of the numerical solution (n, x_n), $n = 0, 1, \ldots, 30$ of the difference equation $x_{n+1} = 2.8x_n(1 - x_n)$ with $x_0 = 0.2$.

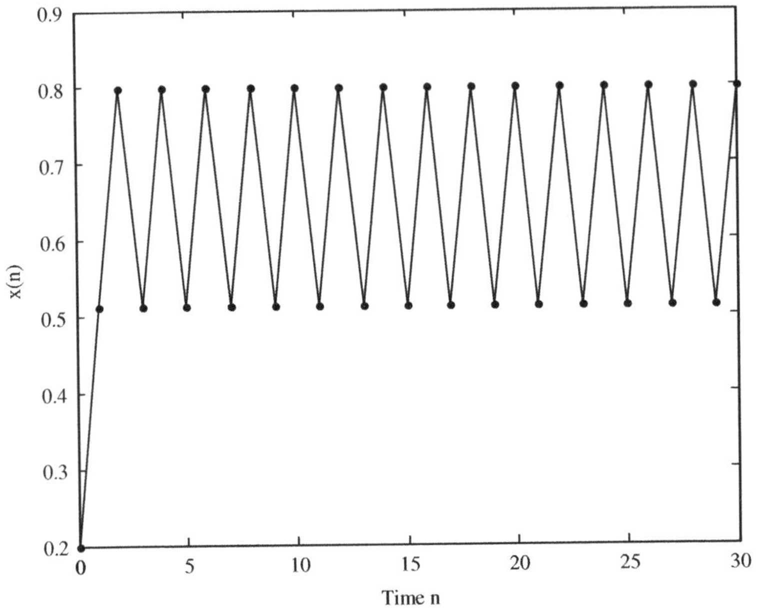

FIGURE 2.10. Graph of the numerical solution (n, x_n), $n= 0, 1, \ldots, 30$ of the difference equation $x_{n+1}=3.2x_n(1-x_n)$ with $x_0=0.2$.

2.4.2. Cyclic Growth: $2<a\leq 2.5699\ldots$ or $3<\lambda\leq 3.5699\ldots$

Period Two Cycle $2<a\leq 2.4495\ldots$ or $3<\lambda\leq 3.4495\ldots$

Let $a=2.2$ (i.e., $\lambda=3.2$) and $x_0=0.2$. Iterate to find (n, x_n), $n = 0, 1, \ldots, 30$ and graph it. The graph is shown in Figure 2.10. The solution and its graph show that after a period of adjusting, x_n oscillates above and below the limiting value. Then x_n oscillates between two values of approximately 0.513044 and 0.799455. In particular $x_n=0.799455$ for even values of n and $x_n=0.513044$ for odd values of n. This type of oscillation is called **two cycle**, and the solution of the difference equation is called **periodic solution with period 2**.

Period Four Cycle $2.4495\ldots<a\leq 2.56\ldots$ or $3.4495\ldots<\lambda\leq 3.56\ldots$

Let $a=2.52$ (i.e., $\lambda=3.52$) and $x_0=0.2$. Iterate to find (n, x_n), $n = 0, 1, \ldots, 30$ and graph it. The graph is shown in Figure 2.11.

Note that x_n repeats itself every four iterations, with two values above the limiting value and two values below it. In this situation the approximate four values are 0.8795, 0.3731, 0.8233, and 0.5121. This kind of oscillation is called **four cycle**. The solution of the difference equation is called **periodic solution with period 4**. Biologically this means that the specie populations (for example, the gypsy moth) undergoes a four–breeding season cycle.

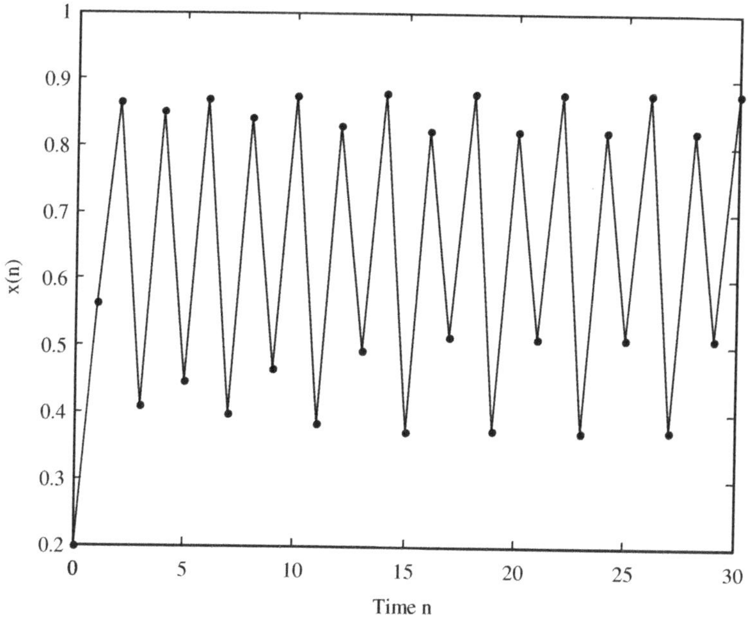

FIGURE 2.11. Graph of the numerical solution (n, x_n), $n = 0, 1, \ldots, 30$ of the difference equation $x_{n+1} = 3.52x_n(1 - x_n)$ with $x_0 = 0.2$.

More Period Doubling 2.56...<a≤2.56994... or 3.56...<λ ≤3.56994...

Beyond $a = 2.56$ (or $\lambda = 3.56$), the period doubles again and again. That is a 4-cycle becomes an 8-cycle and then a 16-cycle and so on. The solutions are periodic, with a 2^n cycle, (8, 16, 32, 64, ...). This is called **periodic doubling**.

2.4.3. Chaotic Growth 2.5699...<a≤3 or 3.5699...<λ≤4

Let $a = 2.7$ (or $\lambda = 3.7$), and $x_0 = 0.2$. Equation 2.35 becomes

$$x_{n+1} = 3.7x_n(1 - x_n).$$

The graph of the numerical solution is shown in Figure 2.12A. Looking carefully at this graph it is clear that the solution does not show any pattern. This is true for any large value of n. The numerical solution (population) does not approach a fixed point or a periodic cycle of any kind. However, the solution does not go to infinity. In fact, it is impossible to predict the long-term behavior of the solution. This type of behavior is called **chaotic**. One feature of chaotic behavior is that the solution closely depends on the initial condition. In other words, a solution is very sensitive to any small change in the initial condition. To explore this feature we will generate the numerical solution of the difference equation with the initial condition $x_0 = 0.201$. The graph is shown in Figure 2.12B. Panel C shows the graphs of two solutions with very close initial

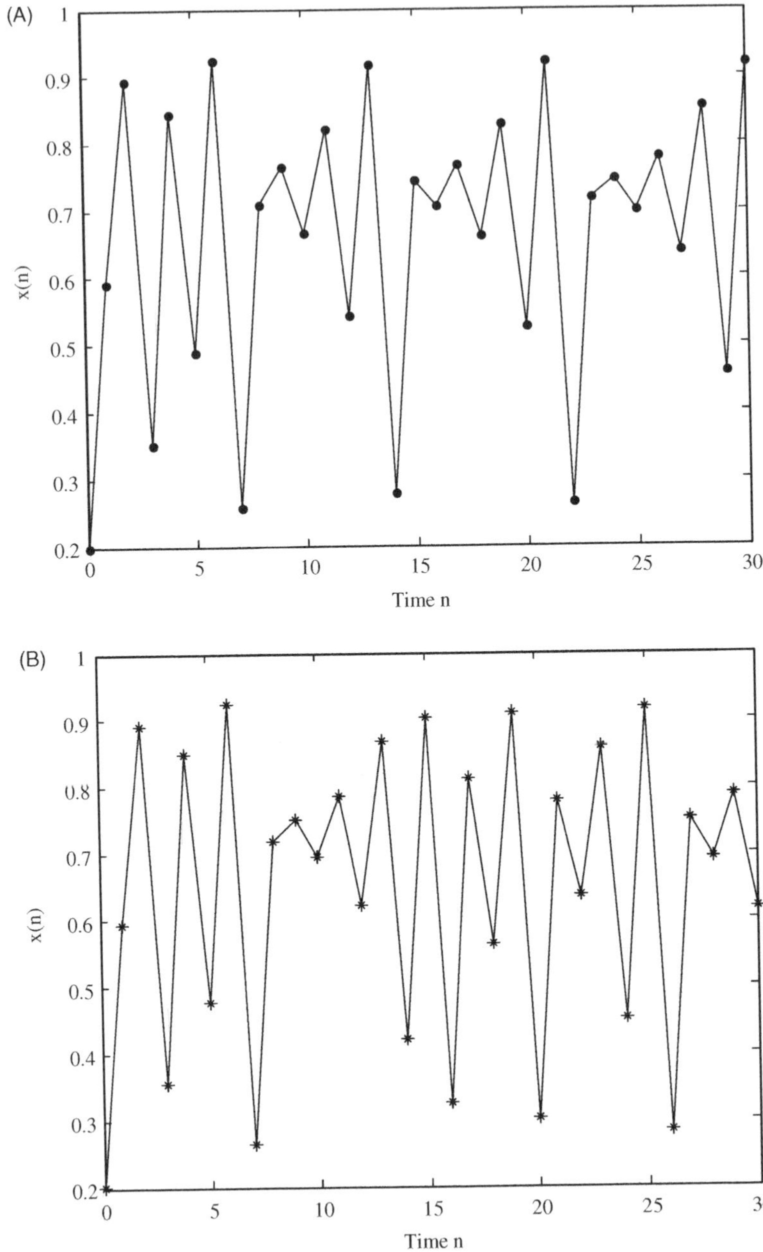

FIGURE 2.12. Graphs of the numerical solution (n, x_n), $n=0, 1, \ldots, 30$ of the difference equation $x_{n+1}=3.7x_n(1-x_n)$. **A.** $x_0=0.2$. **B.** $x_0=0.201$. **C.** $x_0=0.2$ and $x_0=0.201$. **D.** Two graphs of the numerical solution (n, x_n), $n=0, 1, \ldots, 30$ of the difference equation $x_{n+1}=3.8x_n(1-x_n)$, with $x_0=0.4$ and $x_0=0.4001$.

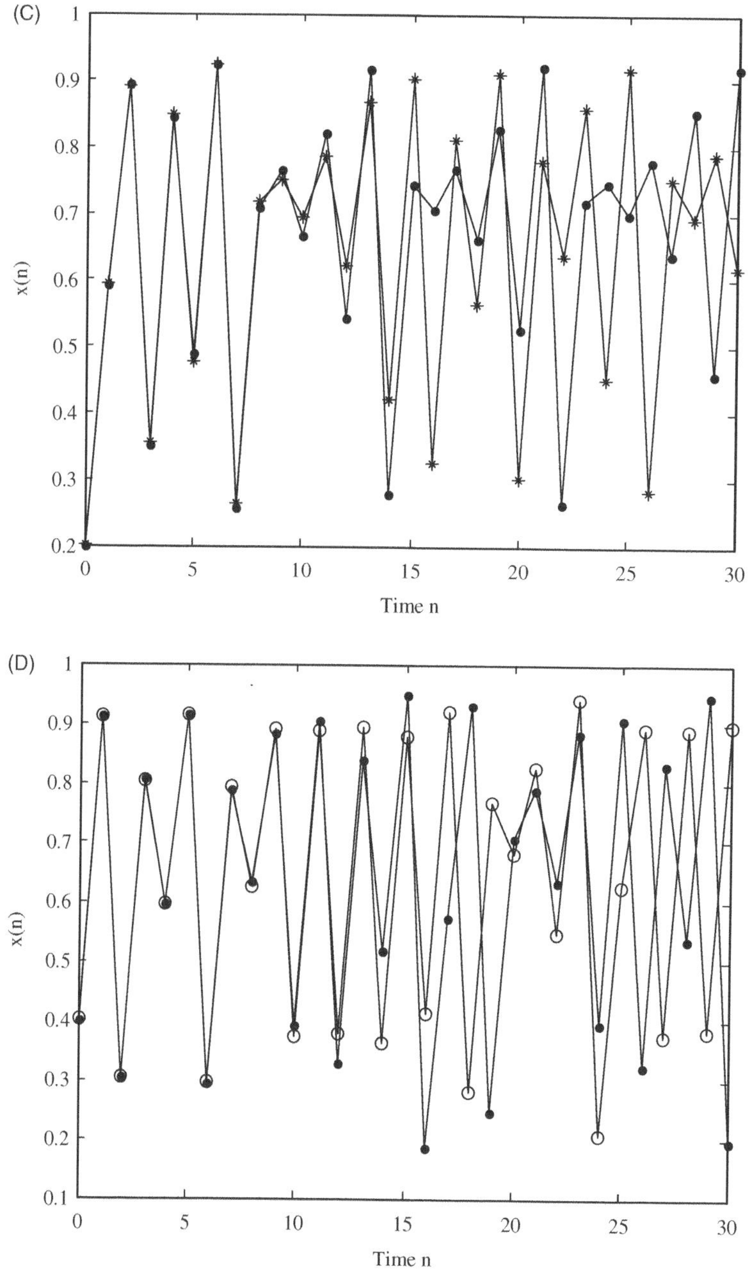

FIGURE 2.12. *(Continued)*

conditions $x_0=0.2$ and $x_0=0.201$. Comparing the graphs in Figure 2.12C, we notice that for a short interval of time (in this case $n = 0, 1, 2, \ldots, 7$), the solutions are almost the same and then they begin to behave differently. Figure 12.12D shows the graphs of the solutions of the difference equation

$$x_{n+1} = 3.8x_n(1 - x_n),$$

with the initial values $x_0=0.4$ and $x_0=0.4001$. Although the initial values are very close the solutions exhibit different behavior.

In general, a system that exhibits irregular oscillation and sensitivity dependence on initial condition is called **chaotic system** and the behavior of the system is called **chaos**.

Note that in this discussion, the range of λ was $1 < \lambda \leq 4$. Let's investigate the quadratic map 2.35, where $\lambda > 4$. Figure 2.13 shows the graph of

$$x_{n+1} = 4.2x_n(1 - x_n), \quad x_0 = 0.2.$$

The scaled population x_n becomes negative at $n \geq 7$.

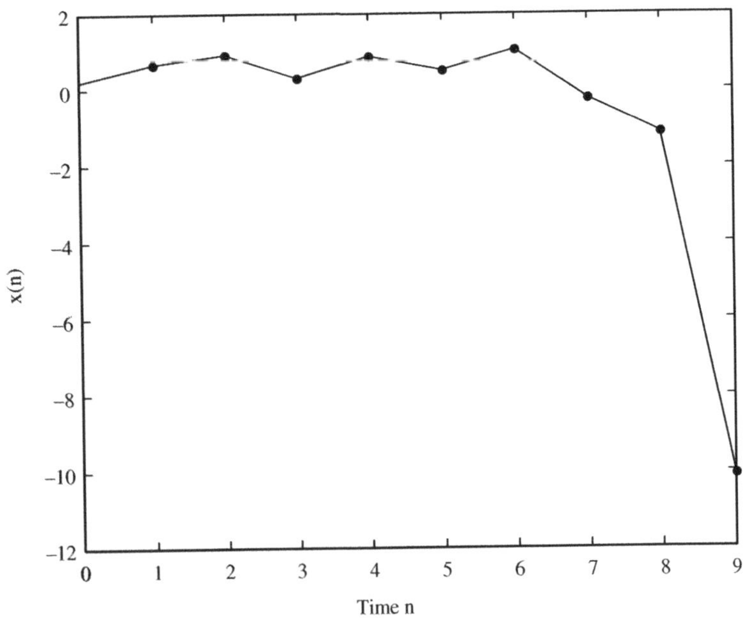

FIGURE 2.13. Graph of the numerical solution (n, x_n), $n=0, 1, \ldots, 30$ of the difference equation $x_{n+1}=4.2x_n(1-x_n)$ with $x_0=0.2$.

2.4.4. Exercises 2.4

In Exercises 1–6 consider species modeled by the logistic difference equation 2.28, $P_{n+1} = P_n + aP_n\left(1 - \frac{P_n}{L}\right)$, where P_n is the population after n time periods, a is the unrestricted growth rate per period, and L is the limiting value (or the carrying capacity) of the environment. Let $L = 1000$. Use MATLAB to do the following:

A. Fix the initial value P_0 (e.g., $P_0 = 200$) and for different values of a create numerical solutions and its graphs. Describe the behavior of the system.

B. Determine if the system is sensitive to the initial values by observing the numerical solutions and its graphs for systems with fixed a and slight change in the initial values (e.g., $P_0 = 200$, 201, and 199).

C. Fix the value of a and change the initial values of P_0 (e.g., $P_0 = 100, 400, 600$) to determine whether the behavior of the system depends on the initial values.

1. Choose a such that $0 < a < 1$.

2. Choose a such that $1 < a \le 2$.

3. Choose a such that $2 < a \le 2.4495$.

4. Choose a such that $2.4495 < a \le 2.56$.

5. Choose a such that $2.56 < a \le 2.56994$.

6. Choose a such that $2.56994 < a \le 3$.

7. Consider species modeled by the difference equation 2.28, $P_{n+1} = P_n + aP_n\left(1 - \frac{P_n}{L}\right)$, with $a > 3$ (e.g., $a = 3.1$). Let $L = 1{,}000$ and $P_0 = 200$. Use MATLAB to graph (n, P_n), $n = 0, 1, \ldots, 10$. Describe the dynamics of the model. Is this model valid for values of the unrestricted growth rate $a > 3$? Explain.

In Exercises 8–10, consider species modeled by the first-order nonlinear difference equation $P_{n+1} = P_n e^{a\left(1 - \frac{P_n}{L}\right)}$, where P_n is the population after n time periods, a is a positive constant, and L is the limiting value (or the carrying capacity). Let $L = 1{,}000$ and $P_0 = 200$. Select the values of a and use MATLAB to answer the questions.

8. Determine whether the system has a stable growth. If it does, determine the range of values of a.

9. Determine whether the system has a cyclic growth. If it does, determine the ranges of values of a where there are two-cycle and four-cycle oscillations.

10. Determine whether the system exhibits chaos. If it does, determine the range of values of a.

In Exercises 11–16 consider species modeled by the difference equation 2.35, $x_{n+1}=\lambda x_n(1-x_n)$, where x_n is the scaled population after n time periods, $\lambda=1+a$, and a is the unrestricted growth rate per period. Use MATLAB to do the following:

A. Fix the initial value x_0 (e.g., $x_0=0.4$) and for different values of λ create numerical solutions and their graphs. Describe the behavior of the system.

B. Determine if the system is sensitive to the initial values by observing the numerical solutions and their graphs for systems with fixed λ and slight change in the initial values (e.g., $x_0 = 0.4, 0.401$, and 0.399).

C. Fix the value of λ and change the initial values of x_0 (e.g., $x_0=0.1$, 0.5, 0.8) to determine whether the behavior of the system depends on the initial values.

11. Choose λ such that $1 < \lambda < 2$.

12. Choose λ such that $2 < \lambda \leq 3$.

13. Choose λ such that $3 < \lambda \leq 3.4495$.

14. Choose λ such that $3.4495 < \lambda \leq 3.56$.

15. Choose λ such that $3.56 < \lambda \leq 3.56994$.

16. Choose λ such that $3.5699 < \lambda \leq 4$.

17. Consider species modeled by the difference equation 2.25, $x_{n+1}=\lambda x_n(1-x_n)$, with $\lambda>4$ (e.g., $\lambda=4.12$). Let $0<x_0<1$ (e.g., $x_0=0.4$). Use MATLAB to graph (n, x_n), $n = 0, 1, \ldots, 10$. Describe the dynamics of the model. Is this model valid for values of the unrestricted growth rate $\lambda>4$? Explain.

CHAPTER 3

MODELING WITH MATRICES

3.1. SYSTEMS OF LINEAR EQUATIONS HAVING UNIQUE SOLUTIONS

In this section I introduce the definition of a system of linear equations and its solution(s). I also introduce a method to find the solution of a system of linear equations when the system has a unique solution. The method is called **Gauss-Jordan elimination**. The Gauss-Jordan elimination method will be introduced in detail in Section 3.2.

Given a system of linear equations, it can be shown that the following operations can be performed to change the system into another system that has the same solution(s) as the original system:

- Any two equations may be interchanged.
- Any equation may be multiplied by a nonzero constant.
- A multiple of an equation may be added to another equation.

These operations are called **elementary transformations**.

Two systems of linear equations are called **equivalent systems**, if one system is derived from the other system by applying elementary transformation(s) on it. Equivalently, the two systems have the same solution. I use the symbol ≈ to indicate that two systems are equivalent.

Explorations of Mathematical Models in Biology with MATLAB®, First Edition. Mazen Shahin.

I develop the Gauss-Jordan method to solve a system of linear equations. Example 3.1 illustrates the method that uses elementary transformations.

Example 3.1

Solve the following system of equations

$$x+2y+z=3 \quad \text{(A)}$$

$$2x-y-3z=1 \quad \text{(B)}$$

$$3x+4y-z=3 \quad \text{(C)}$$

Solution

We use the first equation to eliminate x from the second equation and the third equation. Do the following: Add −2 times the first equation to the second equation. In symbols this would be written as

$$B \leftarrow B+(-2)A.$$

This means that equation B is replaced by B + (−2)A.

Add to the third equation −3 times the first equation. In symbols this is written as

$$C \leftarrow C+(-3)A$$

This yields the equivalent system

$$\begin{aligned} x+2y+\ \ z&=3 \\ 2x-\ \ y-3z&=1 \\ 3x+4y-\ \ z&=3 \end{aligned} \quad \approx \quad \begin{aligned} x+2y+\ \ z&=3. \\ -5y-5z&=-5 \\ -2y-4z&=-6. \end{aligned}$$

$$B \leftarrow B+(-2)A$$
$$C \leftarrow C+(-3)A$$

Next, we want to use the second equation to eliminate y from the first and third equations. To do that, we want to have 1 as a coefficient of y in the second equation. Multiply the second equation by $\frac{-1}{5}$ to obtain

$$\underset{B \leftarrow (-1/5)B}{\approx} \quad \begin{aligned} x+2y+\ \ z&=3. \\ y+\ \ z&=1. \\ -2y-4z&=-6. \end{aligned}$$

To eliminate y from the first equation: A $\leftarrow$ A+(−2)B. To eliminate y from the third equation: C $\leftarrow$ C+(2)B.

$$
\begin{array}{ll}
\approx & x \quad -z=1 \\
\text{A} \leftarrow \text{A}+(-2)\text{B} & y+z=1 \\
\text{C} \leftarrow \text{C}+(2)\text{B} & -2z=-4
\end{array}
$$

Next, get 1 as a coefficient of z in the third equation by multiplying it by $\left(\frac{-1}{2}\right)$ to obtain

$$
\begin{array}{ll}
 & x \quad -z=1. \\
\approx & y+z=1. \\
\text{C} \leftarrow (-1/2)\text{C} & z=2.
\end{array}
$$

Finally, use the third equation to eliminate z from the first and second equations to obtain

$$
\begin{array}{ll}
\approx & x=3. \\
\text{A} \leftarrow \text{A}+\text{C} & y=-1. \\
\text{B} \leftarrow \text{B}+(-1)\text{C} & z=2.
\end{array}
$$

The solution of this system is $x=3$, $y=-1$, $z=2$, which is also a solution to the original system because the two systems are equivalent.

Remark: To check that $x=3$, $y=-1$, and $z=2$ is a solution of the given system, substitute the values in the equations:

A. $\text{LHS} = x+2y+z = 3+2(-1)+2 = 3 = \text{RHS}.$

B. $\text{LHS} = 2x-y-3z = 2(3)-(-1)-3(2) = 1 = \text{RHS}.$

C. $\text{LHS} = 3x+4y-z = 3(3)+4(-1)-(2) = 3 = \text{RHS}.$

3.1.1. Matrices and Systems of Equations

We can avoid writing the variables and equals sign in an original system of linear equations and in the equivalent systems by focusing on the coefficients of the variable and the constant terms. We introduce a more efficient method for solving a system of linear equations. We use a rectangular array of numbers called a **matrix** to describe a system of linear equations. We will discuss matrices in detail in Section 3.3. For now, we will introduce the basic terminology.

A matrix is a rectangular array of numbers. Each number in a matrix is called an **element** of the matrix. Here are some simple matrices:

$$A = \begin{bmatrix} 4 & -2 & 6 \\ 3 & 1 & 9 \\ 2 & 0 & 5 \end{bmatrix}.$$

$$B = \begin{bmatrix} 2 & 4 & -7 \\ 1 & 5 & 3 \end{bmatrix}.$$

$$C = \begin{bmatrix} 4 & 9 \\ -6 & 1 \\ 5 & -2 \end{bmatrix}.$$

$$D = \begin{bmatrix} 2 & 6 & 4 & 7 \end{bmatrix}.$$

$$E = \begin{bmatrix} 4 \\ -7 \\ 1 \end{bmatrix}.$$

A matrix consists of rows and columns. For example, in matrix A, row 1 (symbolized R_1), row 2, and row 3 are $R_1 = \begin{bmatrix} 4 & -2 & 6 \end{bmatrix}$, $R_2 = \begin{bmatrix} 3 & 1 & 9 \end{bmatrix}$, and $R_3 = \begin{bmatrix} 2 & 0 & 5 \end{bmatrix}$. Column 1, column 2, and column 3 are $\begin{bmatrix} 4 \\ 3 \\ 2 \end{bmatrix}$, $\begin{bmatrix} -2 \\ 1 \\ 0 \end{bmatrix}$, and $\begin{bmatrix} 6 \\ 9 \\ 5 \end{bmatrix}$.

A matrix consisting of a single row is called a **row matrix**. Similarly, a **column matrix** is a matrix consisting of a single column. For example, matrix D is a row matrix and matrix E is a column matrix.

The **size** or **dimension** of a matrix consisting of m rows and n columns is denoted by $m \times n$, read "m by n." For example, the size of A is 3×3. B is a 2×3 matrix and C is a 3×2 matrix. The size of the row matrix D is 1×4, and the size of the column matrix E is 3×1.

An element which lies in the ith row and jth column of a matrix A is said to be in location (i, j). A standard notation for this element is a_{ij}, read "a sub i, j." For example in matrix B, element 4 is in location (1, 2), and element 3 is in location (2, 3). In matrix C, element $c_{21} = -6$ and $c_{32} = -2$.

An **identity matrix** is a square matrix with 1's in the diagonal locations of the matrix and zeros in all other locations. I_n is the standard notation for $n \times n$ identity matrix. For example,

$$I_3 = \begin{bmatrix} 1 & 0 & 0 \\ 0 & 1 & 0 \\ 0 & 0 & 1 \end{bmatrix}.$$

$$I_5 = \begin{bmatrix} 1 & 0 & 0 & 0 & 0 \\ 0 & 1 & 0 & 0 & 0 \\ 0 & 0 & 1 & 0 & 0 \\ 0 & 0 & 0 & 1 & 0 \\ 0 & 0 & 0 & 0 & 1 \end{bmatrix}.$$

Two matrices A and B are **equal** ($A=B$) if and only if both matrices have the same size and all the corresponding elements are equal. For example, if

$$\begin{bmatrix} x & 2 \\ 4 & y \\ 6 & 8 \end{bmatrix} = \begin{bmatrix} -7 & 2 \\ 4 & 9 \\ 6 & 8 \end{bmatrix},$$

then $x=-7$ and $y=9$.

Matrices can be used to describe a system of linear equations. There are two matrices associated with a system of linear equations. The **matrix of coefficients** (or **coefficient matrix**) is formed from the coefficients of the variables. The augmented matrix is formed from the coefficients of the variables together with constant terms (the numbers on the right-hand sides of the equations).

Consider the following system of linear equations:

$$\begin{aligned} 2x-6y+3z &= 10. \\ -5x+y+4z &= 7. \\ x+6y-z &= 4. \end{aligned}$$

The coefficient matrix is

$$\begin{bmatrix} 2 & -6 & 3 \\ -5 & 1 & 4 \\ 1 & 6 & -1 \end{bmatrix}$$

and the augmented matrix is

$$\left[\begin{array}{ccc|c} 2 & -6 & 3 & 10 \\ -5 & 1 & 4 & 7 \\ 1 & 6 & -1 & 4 \end{array}\right].$$

Similarly, the augmented matrix

$$\left[\begin{array}{ccc|c} 2 & -6 & 4 & 3 \\ -3 & 0 & 1 & 6 \\ 1 & 5 & 2 & -4 \end{array}\right]$$

represents the following system of linear equations:

$$\begin{aligned} 2x_1 - 6x_2 + 4x_3 &= 3. \\ -3x_1 \qquad\quad + x_3 &= 6. \\ x_1 + 5x_2 + 2x_3 &= -4. \end{aligned}$$

We will use an augmented matrix to represent a system of linear equations. Each equivalent system will also be represented by an augmented matrix.

3.1.2. Elementary Row Operations

Similar to the elementary transformations when we used a system of linear equations, we perform operations on the rows of an augmented matrix representing the system. The following operations can be performed on an augmented matrix:

Any two rows can be interchanged.
Any row can be multiplied by a nonzero constant.
A multiple of a row may be added to another row.

These operations are called **elementary row operations**. If a matrix is derived from another matrix by using elementary row operations, the two matrices are called **row equivalent** or simply **equivalent**.

Now let's see how the Gauss-Jordan method works to solve a system of linear equations by using row operations on the augmented matrix that represents the system.

Example 3.2
Solve the following system of equations:

$$\begin{aligned} 2x_1 + 4x_2 + 6x_3 &= 6. \\ -3x_1 + x_2 + 5x_3 &= -9. \\ 5x_1 + 8x_2 + x_3 &= 5. \end{aligned}$$

Solution

Start by replacing this system with the augmented matrix:

$$\left[\begin{array}{ccc|c} 2 & 4 & 6 & 6 \\ -3 & 1 & 5 & -9 \\ 5 & 8 & 1 & 5 \end{array}\right].$$

Then use the row operations to find the solution to the system.

We need 1 in the location (1, 1), so we multiply R_1 by $\left(\frac{1}{2}\right)$—that is $R_1 \leftarrow \left(\frac{1}{2}\right)R_1$. We obtain the equivalent matrix

$$\left[\begin{array}{ccc|c} 1 & 2 & 3 & 3 \\ -3 & 1 & 5 & -9 \\ 5 & 8 & 1 & 5 \end{array}\right].$$

We need 0 in the location (2, 1) and 0 in the location (3, 1), so we perform the row operations $R_2 \leftarrow R_2 + (3)R_1$, and $R_3 \leftarrow R_3 + (-5)R_1$, respectively, to get

$$\left[\begin{array}{ccc|c} 1 & 2 & 3 & 3 \\ 0 & 7 & 14 & 0 \\ 0 & -2 & -14 & -10 \end{array}\right].$$

Next, to obtain 1 in the location (2, 2), we execute the row operation $R_2 \leftarrow \left(\frac{1}{7}\right)R_2$:

$$\left[\begin{array}{ccc|c} 1 & 2 & 3 & 3 \\ 0 & 1 & 2 & 0 \\ 0 & -2 & -14 & -10 \end{array}\right].$$

We need 0 in the location (1, 2) and 0 in the location (3, 2), so we perform the row operations $R_1 \leftarrow R_1 + (-2)R_2$, and $R_3 \leftarrow R_3 + (2)R_2$, respectively, to get

$$\left[\begin{array}{ccc|c} 1 & 0 & -1 & 3 \\ 0 & 1 & 2 & 0 \\ 0 & 0 & -10 & -10 \end{array}\right].$$

Now we need 1 in the location (3, 3). Let's do $R_3 \leftarrow \left(\frac{-1}{10}\right)R_3$ to get the equivalent matrix

$$\left[\begin{array}{ccc|c} 1 & 0 & -1 & 3 \\ 0 & 1 & 2 & 0 \\ 0 & 0 & 1 & 1 \end{array}\right].$$

Using 1 in the location (3, 3) to get 0 in the locations (1, 3) and (2, 3), we need to perform the row operations $R_1 \leftarrow R_1 + R_3$, and $R_2 \leftarrow R_2 + (-2)R_3$, respectively, to get

$$\left[\begin{array}{ccc|c} 1 & 0 & 0 & 4 \\ 0 & 1 & 0 & -2 \\ 0 & 0 & 1 & 1 \end{array}\right].$$

This last matrix corresponds to the system

$$\begin{aligned} x_1 \qquad\qquad &= 4. \\ x_2 \qquad &= -2. \\ x_3 &= 1. \end{aligned}$$

So the solution is $x_1 = 4$, $x_2 = -2$, and $x_3 = 1$, or in short $(4, -2, 1)$.

Example 3.3

Solve the following system of equations:

$$\begin{aligned} 2x_2 - 4x_3 &= -6. \\ 4x_1 - 8x_2 - 4x_3 &= -4. \\ -3x_1 + 5x_2 + x_3 &= -2. \end{aligned}$$

Solution

The augmented matrix for this system is

$$\left[\begin{array}{ccc|c} 0 & 2 & -4 & -6 \\ 4 & -8 & -4 & -4 \\ -3 & 5 & 1 & -2 \end{array}\right].$$

We need 1 in the location (1, 1), but we have 0. So we will use the row operation $R_1 \leftrightarrow R_2$ to obtain

$$\left[\begin{array}{ccc|c} 4 & -8 & -4 & -4 \\ 0 & 2 & -4 & -6 \\ -3 & 5 & 1 & -2 \end{array}\right]$$

To get 1 in the location (1, 1), let $R_1 \leftarrow \left(\frac{1}{4}\right)R_1$ to obtain

$$\left[\begin{array}{ccc|c} 1 & -2 & -1 & -1 \\ 0 & 2 & -4 & -6 \\ -3 & 5 & 1 & -2 \end{array}\right].$$

We use 1 in the location (1, 1) to get 0 in the locations (2, 1) and (3, 1). Because we have already 0 in the location (2, 1), we need to conduct the row operation $R_3 \leftarrow R_3 + (3)R_1$ to obtain

$$\left[\begin{array}{ccc|c} 1 & -2 & -1 & -1 \\ 0 & 2 & -4 & -6 \\ 0 & -1 & -2 & -5 \end{array}\right].$$

Next, we need 1 in the location (2, 2). Use $R_2 \leftarrow \left(\frac{1}{2}\right)R_2$ to get

$$\left[\begin{array}{ccc|c} 1 & -2 & -1 & -1 \\ 0 & 1 & -2 & -3 \\ 0 & -1 & -2 & -5 \end{array}\right].$$

To get 0 in the locations (1, 1) and (1, 3), let $R_1 \leftarrow R_1 + (2)R_2$ and $R_3 \leftarrow R_3 + R_2$ to get

$$\left[\begin{array}{ccc|c} 1 & 0 & -5 & -7 \\ 0 & 1 & -2 & -3 \\ 0 & 0 & -4 & -8 \end{array}\right].$$

Similarly, we conduct $R_3 \leftarrow \left(\frac{-1}{4}\right)R_3$ to get 1 in the location (3, 3):

$$\left[\begin{array}{ccc|c} 1 & 0 & -5 & -7 \\ 0 & 1 & -2 & -3 \\ 0 & 0 & 1 & 2 \end{array}\right].$$

Finally, we perform the row operations $R_1 \leftarrow R_1 + (5)R_3$ and $R_2 \leftarrow R_2 + (2)R_3$ to get 0 in the locations (1, 3) and (2, 3), respectively:

$$\left[\begin{array}{ccc|c} 1 & 0 & 0 & 3 \\ 0 & 1 & 0 & 1 \\ 0 & 0 & 1 & 2 \end{array}\right].$$

This matrix corresponds to the system

$$\begin{aligned} x_1 \qquad\qquad &= 3. \\ x_2 \qquad &= 1. \\ x_3 &= 2. \end{aligned}$$

Therefore, the system of equations has the unique solution $x_1 = 3$, $x_2 = 1$, and $x_3 = 2$.

Model 3.1: Mixture Problem

A biologist has two solutions of 30% and 70% acid. If she needs 60L of 45% acid, how many liters of each solution should the biologist mix?

Discussion

Let x and y be the number of liters of 30% acid and 70% acid solutions, respectively. Because we need 60L of 45% acid solution, we have

$$x + y = 60.$$

The amount of acid in x L of 30% acid solution is $0.3x$ and the amount of acid in y L of 70% acid solution is $0.7y$. Because the amount of acid in 60L of 45% acid is $0.45(60) = 27$, we have

$$0.3x + 0.7y = 27.$$

The augmented matrix of this system of two linear equations in two variables is

$$\begin{bmatrix} 1 & 1 & 60 \\ 0.3 & 0.7 & 27 \end{bmatrix}.$$

Applying the Gauss-Jordan method we get

$$\begin{bmatrix} 1 & 0 & 37.5 \\ 0 & 1 & 22.5 \end{bmatrix}.$$

Consequently $x = 37.5$ L and $y = 22.5$ L.

Model 3.2: Nutrition

A dietitian plans to provide a patient with a meal that has 70 g protein, 100 g carbohydrates, and 820 mg calcium. The available food is fish, vegetable, and energy drinks. Each serving of fish contains 28 g protein, 36 g carbohydrates, and 207 mg calcium. Each serving of vegetable contains 6 g protein, 36 g carbohydrates, and 12 mg calcium. Each serving of the energy drink contains 11 g protein, 10 g carbohydrates, and 400 mg calcium. Determine the amounts of fish, vegetable, and energy drink needed to meet the protein, carbohydrates, and calcium requirements.

Discussion

The nutrition information in this model can be summarized in the following table

Nutrient	Fish	Vegetable	Energy Drink	Required Amount
Proteins (g)	28	6	11	70
Carbohydrates (g)	36	36	10	100
Calcium (mg)	207	12	400	820

Let x_1, x_2, and x_3 be the number of servings of fish, vegetable, and energy drink, respectively. The amount of protein from x_1 servings of fish, x_2 servings of vegetable, and x_3 servings of energy drink are $28x_1$, $6x_2$, and $11x_3$, respectively. The total amount of protein is $28x_1 + 6x_2 + 11x_3$. Because the patient's meal should contain 70 g proteins, the protein equation is

$$28x_1 + 6x_2 + 11x_3 = 70.$$

Similarly the carbohydrates and energy drink equations are

$$36x_1 + 36x_2 + 10x_3 = 100$$

and

$$207x_1 + 12x_2 + 400x_3 = 820,$$

respectively. The augmented matrix of these linear equations is

$$\begin{bmatrix} 28 & 6 & 11 & 70 \\ 36 & 36 & 10 & 100 \\ 207 & 12 & 400 & 820 \end{bmatrix},$$

and the reduced echelon form is

$$\begin{bmatrix} 1 & 0 & 0 & 2 \\ 0 & 1 & 0 & 0.5 \\ 0 & 0 & 1 & 1 \end{bmatrix}.$$

Consequently, $x_1=2$, $x_2=0.5$, and $x_3=1$. Therefore, the patient's food should contain 2 servings of fish, 0.5 serving of vegetables, and 1 serving of energy drink.

3.1.3. Introduction to Matrices in MATLAB

To enter the matrix

$$A=\begin{bmatrix} 2 & -6 & 3 \\ -5 & 1 & 4 \\ 1 & 6 & -1 \end{bmatrix},$$

at the MATLAB prompt >>type

```
>> A=[2 -6 3; -5 1 4; 1 6 -1]
```

and press **Enter**. The MATLAB outcome is

```
A =
    2 -6  3
   -5  1  4
    1  6 -1
```

Remarks:

- A matrix must be included in brackets, [and].
- The matrix elements are entered by rows.
- The elements of each row are separated by blanks or commas.
- The rows are separated by semicolons or Enter.

For example matrix *A* could be entered in one of these additional forms

```
A=[2, -6, 3; -5, 1, 4; 1, 6, -1]
```

or

```
A=[2 -6  3
   -5  1  4
    1  6 -1]
```

Let's examine two simple methods for using MATLAB to find the unique solution of a system of linear equations such as the systems studied in this section.

For example, to find the solution of the system of linear equations in Example 3.2,

$$2x_1 + 4x_2 + 6x_3 = 6.$$
$$-3x_1 + x_2 + 5x_3 = -9.$$
$$5x_1 + 8x_2 + x_3 = 5.$$

Method 1

1. Enter the augmented matrix of the system and give it a name, say M.
2. Enter the MATLAB function **rref(M)**.

We have

```
>> M=[2 4 6 6; -3 1 5 -9; 5 8 1 5];
>> rref(M)

ans =
   1 0 0  4
   0 1 0 -2
   0 0 1  1
```

We conclude that $x_1=4$, $x_2=-2$, and $x_3=1$. Note that **rref** is a MATLAB predefined function, and it will be introduced in detail in the next section.

Method 2

- Enter the coefficient matrix and name it, say A—that is $A = \begin{bmatrix} 2 & 4 & 6 \\ -3 & 1 & 5 \\ 5 & 8 & 1 \end{bmatrix}$.
- Enter the column matrix of the constants and name it, say B—that is $B = \begin{bmatrix} 6 \\ -9 \\ 5 \end{bmatrix}$.
- Enter the command A\B. The output will be a column matrix of the solutions.

We have

```
>> A=[2 4 6; -3 1 5; 5 8 1];
>> B=[6; -9; 5];
>> A\B
ans =
    4
   -2
    1
```

Thus $x_1=4$, $x_2=-2$, and $x_3=1$.

3.1.4. Exercises 3.1

1. Find the point of the intersection of the lines $-x_1+3x_2=3$ and $2x_1-x_2=4$.
2. Find the point of the intersection of the lines $x_1-2x_2=-4$ and $x_1+x_2=-1$.

Each of the systems in Exercises 3–7 has a unique solution. Solve each system by using the elementary row operations (Gauss-Jordan method).

3. $$\begin{aligned} x_1+2x_2+x_3&=-3\\ -2x_1+2x_2+4x_3&=-6\\ 3x_1+2x_2-2x_3&=-2 \end{aligned}$$

4. $$\begin{aligned} x_2-3x_3&=1\\ 2x_1+2x_2-2x_3&=2\\ -3x_1+x_2&=10 \end{aligned}$$

5. $$\begin{aligned} 2x_1-4x_2-2x_3&=-2\\ x_2-2x_3&=-3\\ -3x_1+5x_2+x_3&=-2 \end{aligned}$$

6. $$\begin{aligned} x_1+2x_2-2x_3&=8\\ 2x_1-3x_2+3x_3&=2\\ -x_1+2x_2+4x_3&=-12 \end{aligned}$$

7. $$\begin{aligned} x_1+x_2+4x_3&=2\\ -2x_1-x_2+2x_3&=1\\ 3x_1+2x_2-2x_3&=-3 \end{aligned}$$

8. A lab technician has two solutions of 40% and 70% acid. If she needs 50L of 56% acid, how many liters of each solution should the technician mix?
9. How should a technician should mix a 60% sulfuric acid solution with 35% sulfuric acid solution to obtain 16L of 43% sulfuric acid?
10. A dietitian is planning a meal for a patient around three foods: I, II, and III. The required percents of proteins, carbohydrates, and iron contained in each ounce of the three foods are shown in the following table:

Nutrient	Food I (%)	Food II (%)	Food III (%)
Proteins	8	12	7
Carbohydrates	6	12	10
Iron	7	13	8

How much of each food should the dietitian use to provide 100% of the needed protein, carbohydrates, and iron? Use MATLAB to determine how many ounces of each food the dietitian should include in the meal for the patient to have exactly 100% of the required protein, carbohydrates and iron.

11. Use MATLAB to show that there is no meaningful solution to the Exercise 10 if the percentage of proteins, carbohydrates, and iron are as shown in the following table:

Nutrient	Food I (%)	Food II (%)	Food III (%)
Proteins	8	12	7
Carbohydrates	6	12	10
Iron	7	13	10

12. A nutritionist is planning a meal for a patient around three foods: I, II, and III. The amount of calcium, iron, and vitamin C in Food I are 22, 1, and 5 mg/oz., respectively. The amount of calcium, iron, and vitamin C in Food II are 28, 2, and 3 mg/oz., respectively; and in Food III, 24, 1, and 4 mg/oz., respectively. The meal should contain 378 mg calcium, 22 mg iron, and 58 mg vitamin C. Use MATLAB to determine how much of each food the nutritionist should use to meet the minimum requirement of calcium, iron, and vitamin C.

3.2. THE GAUSS-JORDAN ELIMINATION METHOD WITH MODELS

In the previous section we discussed the Gauss-Jordan method in solving a system of linear equations, where the number of equations and the number of variables were the same. In addition, the system had a unique solution. In general, a system of linear equations may have one solution, many solutions, or no solution, and the number of equations does not necessarily equal the number of variables.

In this section we discuss the Gauss-Jordan method for general systems of linear equations, where the number of equations may differ from the number of variables and the system may have solution(s) or no solution.

We start with the augmented matrix that represents the system of linear equations. Then we perform a sequence of elementary row operations that transforms the augmented matrix into a simple matrix. This simple matrix can be easily interpreted to determine the existence of a solution of the system. This simpler matrix is called the **reduced echelon form**.

3.2.1. Reduced Echelon Form

A matrix is in reduced echelon form if all the following conditions are satisfied:

A. All rows consisting of allnl zeros are grouped at the bottom of the matrix.

B. The first nonzero element in each row is 1. This element is called the **leading 1**.

C. The leading 1 in any row is to the right of the leading 1 in the previous row.
D. All other entries in the column containing the leading 1 of a row are zero.

For example, the following matrices are all in reduced echelon form. Check the conditions.

$$\begin{bmatrix} 1 & 0 & 0 & 6 \\ 0 & 1 & 0 & -2 \\ 0 & 0 & 1 & 4 \end{bmatrix},$$

$$\begin{bmatrix} 1 & 8 & 0 & 7 \\ 0 & 0 & 1 & 3 \\ 0 & 0 & 0 & 0 \end{bmatrix}$$

$$\begin{bmatrix} 1 & 9 & 0 & 0 \\ 0 & 0 & 1 & 0 \\ 0 & 0 & 0 & 1 \end{bmatrix}$$

$$\begin{bmatrix} 1 & 4 & -3 & 6 \\ 0 & 0 & 0 & 0 \\ 0 & 0 & 0 & 0 \end{bmatrix}$$

$$\begin{bmatrix} 1 & 4 & 2 & 0 & -4 \\ 0 & 0 & 0 & 1 & 9 \\ 0 & 0 & 0 & 0 & 0 \end{bmatrix}.$$

None of the following matrices are in reduced echelon form.

$\begin{bmatrix} 1 & 4 & 0 & 2 \\ 0 & 0 & 0 & 0 \\ 0 & 0 & 1 & 6 \end{bmatrix}$, A row consisting of zeros (the second row) is not at the bottom of the matrix.

$\begin{bmatrix} 1 & 0 & 0 & 6 \\ 0 & 4 & 0 & 9 \\ 0 & 0 & 1 & 7 \end{bmatrix}$, The first nonzero element in row 2 is not 1.

$\begin{bmatrix} 0 & 0 & 0 & 6 \\ 0 & 0 & 1 & 9 \\ 0 & 1 & 0 & 2 \end{bmatrix}$, The leading 1 in row 3 is not to the right of the leading 1 in the previous row.

$$\begin{bmatrix} 1 & 4 & 0 & 6 \\ 0 & 1 & 0 & 5 \\ 0 & 0 & 1 & 7 \end{bmatrix}.$$ The element above the leading 1 in row 2 is not 0.

We use a sequence of row operations on the augmented matrix of a system to obtain the reduced echelon form. The reduced echelon form of an augmented matrix is obtained by first creating the leading 1 in row 1, then zeros below the leading 1. Then create the leading 1 in row 2 and zeros above and below it. Continue in a similar manner and group rows of zeros at the bottom of the matrix. Note that it may be necessary to interchange two rows in order to satisfy condition C for reduced echelon form conditions.

Example 3.4

Use the Gauss-Jordan elimination method to find the reduced echelon form of the following matrix:

$$\begin{bmatrix} 2 & 4 & -2 & 4 & 6 \\ -1 & -2 & 1 & -6 & 5 \\ 3 & 6 & -1 & 4 & 5 \end{bmatrix}.$$

Solution

The elementary row operation $R_1 \leftarrow (\frac{1}{2})R_1$ creates 1 in the location (1, 1),

$$\begin{bmatrix} 1 & 2 & -1 & 2 & 3 \\ -1 & -2 & 1 & -6 & 5 \\ 3 & 6 & -1 & 4 & 5 \end{bmatrix}.$$

The operations $R_2 \leftarrow R_2 + R_1$ and $R_3 \leftarrow R_3 + (-3)R_1$ create zeros in the locations (2, 1) and (3, 1):

$$\begin{bmatrix} 1 & 2 & -1 & 2 & 3 \\ 0 & 0 & 0 & -4 & 8 \\ 0 & 0 & 2 & -2 & -4 \end{bmatrix}.$$

Because we need a leading 1 in row 2 and at the same time must satisfy condition 3 of the reduced echelon form, we perform the operation $R_2 \leftrightarrow R_3$, to obtain

$$\begin{bmatrix} 1 & 2 & -1 & 2 & 3 \\ 0 & 0 & 2 & -2 & -4 \\ 0 & 0 & 0 & -4 & 8 \end{bmatrix}.$$

Now let $R_2 \leftarrow \left(\frac{1}{2}\right)R_2$ to get leading 1 in the location (2, 3):

$$\begin{bmatrix} 1 & 2 & -1 & 2 & 3 \\ 0 & 0 & 1 & -1 & -2 \\ 0 & 0 & 0 & -4 & 8 \end{bmatrix}.$$

The elementary row operation $R_1 \leftarrow R_1 + R_2$ creates 0 in the location (1, 3):

$$\begin{bmatrix} 1 & 2 & 0 & 1 & 1 \\ 0 & 0 & 1 & -1 & -2 \\ 0 & 0 & 0 & -4 & 8 \end{bmatrix}.$$

To get the leading 1 in row 3, we perform $R_3 \leftarrow \left(\frac{-1}{4}\right)R_3$:

$$\begin{bmatrix} 1 & 2 & 0 & 1 & 1 \\ 0 & 0 & 1 & -1 & -2 \\ 0 & 0 & 0 & 1 & -2 \end{bmatrix}.$$

To get zeros in the locations (1, 4) and (2, 4), we perform $R_1 \leftarrow R_1 + (-1)R_3$ and $R_2 \leftarrow R_2 + R_3$:

$$\begin{bmatrix} 1 & 2 & 0 & 0 & 3 \\ 0 & 0 & 1 & 0 & -4 \\ 0 & 0 & 0 & 1 & -2 \end{bmatrix}.$$

This matrix is the reduced echelon form of the given matrix.

3.2.2. Reduced Echelon Form of a Matrix in MATLAB

Let A be a matrix. The MATLAB function **rref(A)** returns the reduced echelon form of the matrix A. The acronym **rref** stands for *row reduced echelon form* of a matrix. For example, to obtain the reduced echelon form of the matrix in Example 3.4,

```
>> A=[2 4 -2 4 6; -1 -2 1 -6 5; 3 6 -1 4 5]
A=
   2  4 -2  4 6
  -1 -2  1 -6 5
   3  6 -1  4 5
>> rref(A)
```

```
ans =
   1 2 0 0  3
   0 0 1 0 -4
   0 0 0 1 -2
```

Example 3.5

Solve, if possible, the following system of equations:

$$2x_1+4x_2+8x_3=-6.$$
$$-3x_1+x_2+9x_3=-5.$$
$$x_1-2x_2-8x_3=5.$$

Solution

We start with the augmented matrix and use Gauss-Jordan elimination method to obtain the reduced echelon form:

$$\begin{bmatrix} 2 & 4 & 8 & -6 \\ -3 & 1 & 9 & -5 \\ 1 & -2 & -8 & 5 \end{bmatrix}.$$

To get a leading 1 in the location (1, 1), let $R_1 \leftarrow (\frac{1}{2})R_1$:

$$\begin{bmatrix} 1 & 2 & 4 & -3 \\ -3 & 1 & 9 & -5 \\ 1 & -2 & -8 & 5 \end{bmatrix}.$$

To get zeros in locations (1, 2) and (1, 3), let $R_2 \leftarrow R_2+(3)R_1$ and $R_3 \leftarrow R_3+(-1)R_1$:

$$\begin{bmatrix} 1 & 2 & 4 & -3 \\ 0 & 7 & 21 & -14 \\ 0 & -4 & -12 & 8 \end{bmatrix}.$$

We need a leading 1 in row 2, so we perform $R_2 \leftarrow (\frac{1}{7})R_2$:

$$\begin{bmatrix} 1 & 2 & 4 & -3 \\ 0 & 1 & 3 & -2 \\ 0 & -4 & -12 & 8 \end{bmatrix}.$$

To get zeros in the locations (1, 2) and (3, 2), we perform $R_1 \leftarrow R_1 + (-2)R_2$ and $R_3 \leftarrow R_3 + (4)R_2$:

$$\begin{bmatrix} 1 & 0 & -2 & 1 \\ 0 & 1 & 3 & -2 \\ 0 & 0 & 0 & 0 \end{bmatrix}.$$

This is the reduced echelon form of the augmented matrix. The corresponding system of equations is

$$\begin{aligned} x_1 \qquad - 2x_3 &= 1. \\ x_2 + 3x_3 &= -2. \end{aligned}$$

We have two linear equations and three unknowns. This indicates that there is no unique solution and the system has many solutions. The leftmost nonzero element in each equation is called the **leading variable** (the leading 1 in the reduced echelon matrix), and the remaining variables are called **free variables**. For example, x_1 is the leading variable and x_3 is the free variable in the first equation, whereas in the second equation, x_2 is the leading variable and x_3 is the free variable. To obtain an explicit form for the solutions of the system, write the leading variable in terms of free variables:

$$\begin{aligned} x_1 &= 2x_3 + 1 \\ x_2 &= -3x_3 - 2. \end{aligned}$$

Because x_3 is a free variable, it can be assigned any real number. Assign the arbitrary value t to the free variable x_3. The t is called a **parameter**. In terms of the parameter t, the **general solution** to the system has the following form:

$$\begin{aligned} x_1 &= 2t + 1 \\ x_2 &= -3t - 2. \\ x_3 &= t \end{aligned}$$

A particular solution can be obtained by assigning a specific value for the parameter t. For example,

When $t=0$, the solution is $x_1=1$, $x_2=-2$, $x_3=0$.

When $t=1$, the solution is $x_1=3$, $x_2=-5$, $x_3=1$.

When $t=-2$, the solution is $x_1=-3$, $x_2=4$, $x_3=-2$.

When $t=10$, the solution is $x_1=21$, $x_2=-32$, $x_3=10$.

To use MATLAB to obtain the reduced echelon form of the matrix considered in this example, enter the matrix and name it, say A, then use the function **rref(A)**. Here are the MATLAB commands and output:

```
>> A=[2 4 8 -6; -3 1 9 -5; 1 -2 -8 5]
A=
    2   4   8  -6
   -3   1   9  -5
    1  -2  -8   5

>> rref(A)
ans=
   1 0 -2  1
   0 1  3 -2
   0 0  0  0
```

Example 3.6

Solve, if possible, the following system of equations:

$$\begin{aligned} x_1 - x_2 - x_3 - 5x_4 &= 3. \\ -2x_1 + x_2 + x_3 + 6x_4 &= -5. \\ -3x_1 + 2x_2 + 4x_3 + 11x_4 &= -8. \end{aligned}$$

Solution

We start with the augmented matrix and transform it to the reduced echelon form. The augmented matrix is

$$\begin{bmatrix} 1 & -1 & -1 & -5 & 3 \\ -2 & 1 & 3 & 6 & -5 \\ -3 & 2 & 4 & 11 & -8 \end{bmatrix}.$$

We use the leading 1 in row 1 to get 0 in the locations (2, 1) and (3, 1) by performing the operations, $R_2 \leftarrow R_2 + (2)R_1$, and $R_3 \leftarrow R_3 + (3)R_1$, respectively:

$$\begin{bmatrix} 1 & -1 & -1 & -5 & 3 \\ 0 & -1 & 1 & -4 & 1 \\ 0 & -1 & 1 & -4 & 1 \end{bmatrix}.$$

To get a leading 1 in the location (2, 2), we perform $R_2 \leftarrow (-1)R_2$, to get

$$\begin{bmatrix} 1 & -1 & -1 & -5 & 3 \\ 0 & 1 & -1 & 4 & -1 \\ 0 & -1 & 1 & -4 & 1 \end{bmatrix}.$$

Now use the leading 1 in row 2 to obtain 0 in the locations (1, 2) and (3, 2) by executing the following elementary row operations, respectively, $R_1 \leftarrow R_1 + R_2$ and $R_3 \leftarrow R_3 + R_2$ to obtain

$$\begin{bmatrix} 1 & 0 & -2 & -1 & 2 \\ 0 & 1 & -1 & 4 & -1 \\ 0 & 0 & 0 & 0 & 0 \end{bmatrix}.$$

This reduced echelon form represents the following system of equations:

$$\begin{aligned} x_1 \qquad -2x_3 - \ x_4 &= 2. \\ x_2 - \ x_3 + 4x_4 &= -1. \end{aligned}$$

The leading variables are x_1 and x_2, and the free variables are x_3 and x_4. Expressing the leading variables in terms of the free variables, we get

$$\begin{aligned} x_1 &= 2x_3 + x_4 + 2. \\ x_2 &= \ x_3 - 4x_4 - 1. \end{aligned}$$

Let's assign the arbitrary values (parameters) r to x_3 and t to x_4. In terms of these parameters, the general solution is

$$\begin{aligned} x_1 &= 2r + t + 2. \\ x_2 &= r - 4t - 1. \\ x_3 &= r. \\ x_4 &= t. \end{aligned}$$

Given particular values of the parameters r and t, we get a specific solution. For example, if we set $r=1$, and $t=2$, then the solution of the system of equations is $x_1=6$, $x_2=-8$, $x_3=1$, $x_4=2$. Verify that $(6,-8,1,2)$ is a solution of the given system of equations.

Here is MATLAB's calculation of reduced echelon form of the matrix discussed in this example:

```
>> A=[1 -1 -1 -5 3; -2 1 3 6 -5; -3 2 4 11 -8]
A=
    1 -1 -1 -5  3
   -2  1  3  6 -5
   -3  2  4 11 -8
```

```
>> rref(A)
ans =
   1 0 -2 -1  2
   0 1 -1  4 -1
   0 0  0  0  0
```

A system of linear equations may not have a solution. The following example illustrates how to use reduced echelon form of an augmented matrix to determine that a system of equations has no solution.

Example 3.7

Solve, if possible, the following system of equations:

$$\begin{aligned} x_1 + 2x_2 + x_3 &= 7. \\ -x_1 - x_2 + x_3 &= -4. \\ -3x_1 - 7x_2 - 5x_3 &= 1. \end{aligned}$$

Solution

As usual, we start with the augmented matrix and use the Gauss-Jordan elimination method to obtain the reduced echelon form. We have

$$\begin{bmatrix} 1 & 2 & 1 & 7 \\ -1 & -1 & 1 & -4 \\ -3 & -7 & -5 & 1 \end{bmatrix}.$$

Using the leading 1 in the location (1, 1) to get zeros in the locations (2, 1) and (3, 1), set $R_2 \leftarrow R_2 + R_1$ and $R_3 \leftarrow R_3 + (3)R_1$ to obtain

$$\begin{bmatrix} 1 & 2 & 1 & 7 \\ 0 & 1 & 2 & 3 \\ 0 & -1 & -2 & 22 \end{bmatrix}.$$

Use the leading 1 in the location (2, 2) to get zeros in the locations (1, 2) and (3, 2). Set $R_1 \leftarrow R_1 + (-2)R_2$ and $R_3 \leftarrow R_3 + R_2$ to get

$$\begin{bmatrix} 1 & 0 & -3 & 1 \\ 0 & 1 & 2 & 3 \\ 0 & 0 & 0 & 25 \end{bmatrix}.$$

This matrix is not yet the reduced echelon form because a leading 1 can be created in the last row and then zeros can be created above the leading 1. However, we do not need to proceed further. Why? The equivalent system of equations of the last matrix is

$$\begin{aligned} x_1 \qquad -3x_3 &= 1. \\ x_2 + 2x_3 &= 3. \\ 0x_1 + 0x_2 + 0x_3 &= 25. \end{aligned}$$

There are no values for x_1, x_2, and x_3 that satisfy the last equation. Therefore, the system has no solution.

Here is MATLAB's calculation of reduced echelon form of the matrix discussed in this example:

```
>> A=[1 2 1 7; -1 -1 1 -4; -3 -7 -5 1]
A=
   1  2  1  7
  -1 -1  1 -4
  -3 -7 -5  1
>> rref(A)
ans=

  1 0 -3 0
  0 1  2 0
  0 0  0 1
```

3.2.3. Homogenous Systems of Linear Equations

A system of linear equations is said to be **homogenous** if the constant terms are zeros. For example, the following systems are homogenous:

$$\begin{aligned} 2x_1 + 4x_2 - 6x_3 &= 0. \\ 5x_2 + 9x_3 &= 0. \\ x_1 - 5x_2 + \ x_3 &= 0. \end{aligned}$$

$$\begin{aligned} 4x_1 - 2x_2 + 9x_3 &= 0. \\ 6x_2 - 2x_3 &= 0. \end{aligned}$$

Obviously, $x_1=0$, $x_2=0$, and $x_3=0$ is a solution for each of these homogenous systems. This solution is called a **trivial solution**. In general, any homogenous system has a trivial solution.

Usually, we are interested in nontrivial solutions of a homogenous system. We will use the Gauss-Jordan method to determine whether a homogenous system has a nontrivial solution(s).

Example 3.8

Solve, if possible, the following system of equations:

$$\begin{aligned} x_1 + x_2 + 2x_3 &= 0. \\ -4x_1 + \quad\quad 8x_3 &= 0. \\ 3x_1 + 2x_2 + 2x_3 &= 0. \end{aligned}$$

Solution

We start with the augmented matrix and use the Gauss-Jordan elimination method to obtain the reduced echelon form. We have

$$\begin{bmatrix} 1 & 1 & 2 & 0 \\ -4 & 0 & 8 & 0 \\ 3 & 2 & 2 & 0 \end{bmatrix}.$$

We use the leading 1 in row 1 to get zeros in the locations (2, 1) and (3, 1) by performing the operations $R_2 \leftarrow R_2 + (4)R_1$ and $R_3 \leftarrow R_3 + (-3)R_1$ to get

$$\begin{bmatrix} 1 & 1 & 2 & 0 \\ 0 & 4 & 16 & 0 \\ 0 & -1 & -4 & 0 \end{bmatrix}.$$

We need 1 in the location (2, 2). Letting $R_2 \leftarrow \left(\frac{1}{4}\right)R_2$, we get

$$\begin{bmatrix} 1 & 1 & 2 & 0 \\ 0 & 1 & 4 & 0 \\ 0 & -1 & -4 & 0 \end{bmatrix}.$$

We perform the operations $R_1 \leftarrow R_1 + (-1)R_2$ and $R_3 \leftarrow R_3 + R_2$ to get zeros in the locations (1, 2) and (3, 2):

$$\begin{bmatrix} 1 & 0 & -2 & 0 \\ 0 & 1 & 4 & 0 \\ 0 & 0 & 0 & 0 \end{bmatrix}.$$

This reduced echelon form corresponds to the following system of equations:

$$\begin{aligned} x_1 \quad\quad - 2x_3 &= 0. \\ x_2 + 4x_3 &= 0. \\ 0 &= 0. \end{aligned}$$

Writing the leading variables in terms of the free variables, we get

$$x_1 = 2x_3.$$
$$x_2 = -4x_3.$$

Letting $x_3 = r$, the general solution of the homogenous system is

$$x_1 = 2r.$$
$$x_2 = -4r.$$
$$x_3 = r.$$

This indicates that the system has an infinite number of solutions. Note that the trivial solution is obtained by choosing $r = 0$.

Model 3.3: Nutrition

A dietitian is planning a meal for a patient around three foods: I, II, and III. The percents of required protein, carbohydrates, and calcium contained in each ounce of the three foods are shown in the following table:

Nutrient	Food I (%)	Food II (%)	Food III (%)
Proteins	12	10	2
Carbohydrates	6	10	8
Calcium	7	5	10

How many ounces of each food should the dietitian include in the meal for a patient to have exactly 100% of the required protein, carbohydrates, and calcium?

Discussion

Let x_1, x_2, and x_3 be the number of ounces of Foods I, II, and III, respectively. Because each ounce of Food I has 12% of required protein, $0.12x_1$ would be the amount of protein in x_1 ounces. Similarly, $0.10x_2$ and $0.02x_3$ are the amount of protein in x_2 and x_3 ounces of Food II and Food III, respectively. The total amount of protein in Food I, II, and III is

$$0.12x_1 + 0.10x_2 + 0.02x_3.$$

Because the total required proteins must be 100%, the protein equation is

$$0.12x_1 + 0.10x_2 + 0.02x_3 = 1$$

or

$$12x_1 + 10x_2 + 2x_3 = 100.$$

Similarly, the carbohydrates and the calcium equations are as follows:

$$6x_1 + 10x_2 + 8x_3 = 100$$

and

$$7x_1 + 5x_2 + 10x_3 = 100,$$

respectively.

To find x_1, x_2, and x_3 we find the reduced echelon form of the augmented matrix:

$$\begin{bmatrix} 12 & 10 & 2 & 100 \\ 6 & 10 & 8 & 100 \\ 7 & 5 & 10 & 100 \end{bmatrix}.$$

The reduced echelon form is

$$\begin{bmatrix} 1 & 0 & 0 & 5 \\ 0 & 1 & 0 & 3 \\ 0 & 0 & 1 & 5 \end{bmatrix}.$$

Therefore, $x_1 = 5$, $x_2 = 3$, and $x_3 = 5$. The dietitian should provide the patient with 5 oz. of Food I, 3 oz. of Food II, and 5 oz. of Food III.

Let's see if the calcium in Food III was 5% instead of 10%. The model is represented by the following three linear equations:

$$12x_1 + 10x_2 + 2x_3 = 100.$$

$$6x_1 + 10x_2 + 8x_3 = 100.$$

$$7x_1 + 5x_2 + 5x_3 = 100.$$

The augmented matrix of the system is

$$\begin{bmatrix} 12 & 10 & 2 & 100 \\ 6 & 10 & 8 & 100 \\ 7 & 5 & 5 & 100 \end{bmatrix},$$

and the reduced echelon form is

$$\begin{bmatrix} 1 & 0 & 0 & 10 \\ 0 & 1 & 0 & -4 \\ 0 & 0 & 1 & 10 \end{bmatrix}.$$

This implies that the solution is $x_1=10$, $x_2=-4$, and $x_3=10$. However, the number of ounces of Food II, x_2, is a negative number, which does not make sense in this context. Consequently, it is not possible to provide exactly 100% of protein, carbohydrates, and calcium from the three foods with that information.

3.2.4. Model 3.4: Allocation of Resources

In this model we study the allocation of limited resources under a set of constrains. Let's study the following situation.

A biologist has three species of bacteria, denoted A, B, and C in her research lab. The bacteria are fed three different foods, denoted I, II, and III. The three species of bacteria consume the three foods each day as follows. Each bacterium from bacteria A consumes 4 units from Food I, 3 units from Food II, and 5 units from Food III. Each bacterium of type B consumes 3, 2, and 3 units from Foods I, II, and III, respectively. Each bacterium of type C consumes 4 units from Food I and 4 units from Food II. If the biologist has 6000 units of Food I, 5000 units of Food II, and 4000 units of Food III every day, how many bacteria of each species can co-exist in the research lab.

Discussion

The information in this situation can be summarized in the following table:

Food	Bacteria A	Bacteria B	Bacteria C
I (units)	4	3	4
II (units)	3	2	4
III (units)	5	3	0

Let x_1, x_2, and x_3 be the number of bacteria A, B, and C, respectively. The number of units of Food I consumed by x_1 bacteria is $4x_1$. The numbers of units of Food I consumed by x_2 and x_3 bacteria are $3x_2$ and $4x_3$, respectively. The total units of Food I consumed by the three species of bacteria is $4x_1+3x_2+4x_3$. Because the number of available units of food I that would be consumed by the three bacteria each day is 6000 units, we have the equation $4x_1+3x_2+4x_3=6{,}000$ for Food I. Similarly, the equations for Foods II and III are $3x_1+2x_2+4x_3=5{,}000$ and $5x_1+3x_2+0x_3=4{,}000$, respectively.

To solve this system of linear equations, we calculate the reduced echelon form of the augmented matrix:

$$\begin{bmatrix} 4 & 3 & 4 & 6000 \\ 3 & 2 & 4 & 5000 \\ 5 & 3 & 0 & 4000 \end{bmatrix},$$

which is

$$\begin{bmatrix} 1 & 0 & 0 & 500 \\ 0 & 1 & 0 & 500 \\ 0 & 0 & 1 & 625 \end{bmatrix}.$$

Consequently, $x_1 = 500$, $x_2 = 500$, and $x_3 = 625$. This means 500 bacteria A, and 500 bacteria B, and 625 bacteria C can co-exist in the research lab.

3.2.5. Model 3.5: Balancing Chemical Equations

Consider the following chemical reaction

$$\text{Ethane} + \text{Oxygen} \rightarrow \text{Carbon Dioxide} + \text{Steam}$$

This reaction can be represented by the chemical equation

$$C_2H_6 + O_2 \rightarrow CO_2 + H_2O$$

The gas ethane (C_2H_6) burns in oxygen (O_2) and produces carbone dioxide (CO_2) and steam (or water (H_2O)).

A chemical reaction is a process that transforms one set of chemical substances, called **reactants**, to another set, called **products**. Note that we use an arrow in chemical equations instead of the equals sign used in mathematics equations.

In this example, the reactants are C_2H_6 and O_2, and the products are CO_2 and H_2O.

The chemical equation is not balanced and needs to be balanced. To balance this chemical equation, find positive integer coefficients of the reactants and products so that the number of atoms per molecule of carbon (C), hydrogen (H), and oxygen (O) in the left side and the right side are equal. We have two reactants and two products. Let x_1, x_2, x_3, and x_4 be the number of molecules of C_2H_6, O_2, CO_2, and H_2O, respectively. Therefore, the chemical equation becomes

$$x_1C_2H_6 + x_2O_2 \rightarrow x_3CO_2 + x_4H_2O$$

We need to compare the numbers of carbon, hydrogen, and oxygen atoms in both sides of the chemical equation:

$$\begin{aligned} \text{C:} \quad & 2x_1 = x_3 \\ \text{H:} \quad & 6x_1 = 2x_4 \\ \text{O:} \quad & 2x_2 = 2x_3 + x_4 \end{aligned}$$

Rewriting these equations in standard form, we get

$$\begin{aligned} 2x_1 \quad - x_3 \quad &= 0. \\ 6x_1 \quad - 2x_4 &= 0. \\ 2x_2 - 2x_3 - x_4 &= 0. \end{aligned}$$

To solve this homogenous system, row reduce the augmented matrix

$$\left[\begin{array}{cccc|c} 2 & 0 & -1 & 0 & 0 \\ 6 & 0 & 0 & -2 & 0 \\ 0 & 2 & -2 & -1 & 0 \end{array}\right].$$

The reduced echelon form is

$$\left[\begin{array}{cccc|c} 1 & 0 & 0 & -1/3 & 0 \\ 0 & 1 & 0 & -7/6 & 0 \\ 0 & 0 & 1 & -2/3 & 0 \end{array}\right].$$

Consequently,

$$x_1 - \frac{1}{3}x_4 = 0.$$

$$x_2 - \frac{7}{6}x_4 = 0.$$

$$x_3 - \frac{2}{3}x_4 = 0.$$

This reduces to

$$x_1 = \frac{1}{3}x_4.$$

$$x_2 = \frac{7}{6}x_4.$$

$$x_3 = \frac{2}{3}x_4.$$

Letting $x_4 = r$, a parameter, the system has many solutions in the form

$$x_1 = \frac{1}{3}r.$$

$$x_2 = \frac{7}{6}r.$$

$$x_3 = \frac{2}{3}r.$$

$$x_4 = r.$$

Because we are looking for integer values of x_1, x_2, x_3, and x_4, the least value for r is 6. We have the solution:

$$x_1 = 2.$$

$$x_2 = 7.$$

$$x_3 = 4.$$

$$x_4 = 6.$$

Consequently, the balanced chemical equation is

$$2C_2H_6 + 7O_2 \rightarrow 4CO_2 + 6H_2O.$$

Check that each side of the equation has 4 atoms of C, 12 atoms of H, and 14 atoms of O.

Note that the MATLAB command **`format rat`** displays the numbers in rational format.

3.2.6. Exercises 3.2

In Exercises 1–6, determine whether the matrices are in reduced echelon form. Give the reason if the matrix is not in reduced echelon form.

1. $\begin{bmatrix} 1 & 0 & 0 & 2 \\ 0 & 1 & 0 & 4 \\ 0 & 0 & 1 & -3 \end{bmatrix}.$

2. $\begin{bmatrix} 1 & 6 & 0 & 8 \\ 0 & 0 & 1 & 2 \\ 0 & 0 & 0 & 0 \end{bmatrix}.$

3. $\begin{bmatrix} 2 & 4 & 0 & 3 \\ 0 & 0 & 0 & 0 \\ 0 & 0 & 1 & 5 \end{bmatrix}.$

4. $\begin{bmatrix} 1 & 0 & 0 & 4 \\ 0 & 0 & 1 & 6 \\ 0 & 1 & 0 & 3 \end{bmatrix}.$

5. $\begin{bmatrix} 1 & 2 & 5 & 0 & 7 \\ 0 & 0 & 0 & 1 & -6 \\ 0 & 0 & 0 & 0 & 0 \end{bmatrix}.$

6. $\begin{bmatrix} 1 & 6 & 0 & 0 & 4 \\ 0 & 0 & 1 & 0 & 3 \\ 0 & 0 & 0 & 1 & -2 \end{bmatrix}.$

In Exercises 7–12, the matrix is the reduced echelon form of the linear system. Determine the solution if the system has a unique solution or if the parametric form of the solutions if the system has many solutions. If the system has no solution, give the reason.

7. $\begin{bmatrix} 1 & 0 & 0 & 3 \\ 0 & 1 & 0 & -2 \\ 0 & 0 & 1 & 4 \end{bmatrix}.$

8. $\begin{bmatrix} 1 & 0 & 2 & -3 \\ 0 & 1 & -3 & 6 \\ 0 & 0 & 0 & 0 \end{bmatrix}.$

9. $\begin{bmatrix} 1 & 0 & -2 & 0 \\ 0 & 1 & 3 & 0 \\ 0 & 0 & 0 & 1 \end{bmatrix}.$

10. $\begin{bmatrix} 1 & 4 & 0 & -5 \\ 0 & 0 & 1 & 3 \\ 0 & 0 & 0 & 0 \end{bmatrix}.$

11. $\begin{bmatrix} 1 & 2 & 0 & 0 & -4 \\ 0 & 0 & 1 & 0 & 3 \\ 0 & 0 & 0 & 1 & -2 \end{bmatrix}.$

12. $\begin{bmatrix} 1 & 0 & 0 & 4 & 2 \\ 0 & 1 & 0 & 5 & -4 \\ 0 & 0 & 1 & 3 & 2 \end{bmatrix}.$

In Exercises 13–16, use the Gauss-Jordan elimination method to find solutions of the linear systems. Verify your solution by using MATLAB's function **rref**.

13.
$$\begin{aligned} x_1 + 2x_2 + x_3 &= -3 \\ 3x_1 + x_2 - 3x_3 &= 1 \\ -4x_1 - x_2 + 5x_3 &= -2 \end{aligned}$$

14.
$$\begin{aligned} x_1 + 2x_2 + 4x_3 &= -3 \\ 3x_1 + x_2 - 3x_3 &= 1 \\ -4x_1 - x_2 + 5x_3 &= -2 \end{aligned}$$

15.
$$\begin{aligned} -2x_1 + 2x_2 + 6x_3 &= -2 \\ 3x_1 + x_2 - x_3 &= 7 \\ x_1 + 3x_2 + 5x_3 &= 5 \end{aligned}$$

16.
$$\begin{aligned} x_1 + x_2 - x_3 &= 2 \\ -2x_1 - x_2 + 4x_3 &= 1 \\ 3x_1 + 4x_2 - x_3 &= -3 \end{aligned}$$

In Exercises 17–19, use the Gauss-Jordan elimination method to find a nontrivial solution of the homogenous systems. Verify your solution by using MATLAB's function **rref**.

17.
$$\begin{aligned} x_1 + 3x_2 + 3x_3 &= 0 \\ -3x_1 - 5x_2 - x_3 &= 0 \\ 2x_1 + x_2 - 4x_3 &= 0 \end{aligned}$$

18.
$$\begin{aligned} x_1 + 3x_2 + 2x_3 &= 0 \\ 2x_1 + x_2 - 6x_3 &= 0 \end{aligned}$$

19.
$$\begin{aligned} x_1 + x_2 + 2x_3 - 5x_4 &= 0 \\ x_2 + 4x_3 - 2x_4 &= 0 \\ 3x_1 + 2x_2 + 2x_3 - 13x_4 &= 0 \end{aligned}$$

20. A linear equation in two variables represents a straight line in two-dimensional (2D) space and a linear equation in three variables represents a plane in three-dimensional (3D) space. In other words, the equation of a plane in a 3D space is in the following form: $ax_1 + bx_2 + cx_3 = d$, where a, b, and c are constants, and not all a, b, and c are zero.

Given a system of three linear equations in three variables, give geometric interpretations for a unique solution, infinite number of solutions, and no solution of the system.

21. A dietitian is planning a meal for a patient around three foods: I, II, and III. The percent of required protein, carbohydrates, and iron contained in each ounce of the three foods is shown in the following table:

Nutrient	Food I (%)	Food II (%)	Food III (%)
Protein	12	6	5
Carbohydrates	8	10	6
Iron	6	4	10

Use MATLAB to determine how many ounces of each food the dietitian should include in the meal for the patient to have exactly 100% of protein, carbohydrates, and iron.

22. Use MATLAB to show that there is no meaningful solution to Exercise 21, if the percentage of proteins, carbohydrates, and iron are summarized as in the following table:

Nutrient	Food I (%)	Food II (%)	Food III (%)
Protein	8	7	8
Carbohydrates	8	10	6
Iron	6	5	9

23. A nutritionist is planning a meal for a patient around three foods: I, II, and III. The amount of calcium, iron, and vitamin C in Food I are 22, 1, and 5 mg/oz., respectively. The amount of calcium, iron, and vitamin C in Food II are 28, 2, and 3 mg/oz., respectively; and in Food III, 24, 1, and 4 mg/oz., respectively. The meal should contain at least 390 mg calcium, 22 mg iron, and 58 mg vitamin C. Use MATLAB to determine how much of each food the nutritionist should use to meet the minimum requirement of calcium, iron, and vitamin C?

24. A biologist has three species of bacteria, denoted A, B, and C, in her research lab. The bacteria are fed three different foods, denoted I, II, and III. The three species of bacteria consume the three foods each day as follows. Each bacterium from bacteria A consumes 2 units from Food I and 5 units from Food III. Each bacterium of type B consumes 3, 5, and 3 units from Foods I, II, and III respectively. Each bacterium of type C consumes 4 units from Food I, 3 units from Food II, and 3 units from Food III. If the biologist has 5500 units of Food I, 4600 units of Food II, and 6600 units of Food III every day, how many bacteria of each species can co-exist in the research lab.

In Exercises 25–28, use matrices and MATLAB to determine the values of x_1, x_2, x_3, and x_4 to balance the chemical equations associated with the given chemical reactions. *Hint*: Use rational format **format rat** to calculate the row reduced echelon form of the coefficient matrix of the homogenous system of equations.

25. Ammonia + oxygen → nitric oxide + water

$$x_1 NH_3 + x_2 O_2 \rightarrow x_3 NO + x_4 H_2O$$

26. Lead + phosphoric acid → hydrogen + lead phosphate

$$x_1 Pb + x_2 H_3PO_4 \rightarrow x_3 H + x_4 Pb_3(PO_4)_2$$

27. Perchloric acid + tetraphosphorus decaoxide → phosphoric acid + dichlorine heptoxide

$$x_1 HClO_4 + x_2 P_4O_{10} \rightarrow x_3 H_3PO_4 + x_4 Cl_2O_7$$

28. Dodecane + oxygen → carbon dioxide + water

$$x_1 C_{12}H_{26} + x_2 O_2 \rightarrow x_3 CO_2 + x_4 H_2O$$

3.3. INTRODUCTION TO MATRICES

In the previous two sections we used matrices in the form of augmented matrices to solve systems of linear equations. In this section we introduce the standard matrix notation and the basic matrix operations such as addition, subtraction, scalar multiplication, multiplication, and power. We will see how a system of linear equations is represented by a matrix equation—that is, an equation involving matrices. We will study the inverse of a square matrix

3.3.1. Some Matrix Notation

A **matrix** is a rectangular array of numbers called **entries** or **elements**. We use capital letters to represent matrices. The size of a matrix with m rows and n columns is written $m \times n$ (read "m by n"). The entry of a matrix A in row i and column j is written as a_{ij}. We call this the (i, j) entry of the matrix A. The matrix A can be represented as

$$A = \begin{bmatrix} a_{11} & a_{12} & \dots & a_{1n} \\ a_{21} & a_{22} & \dots & a_{2n} \\ \vdots & & & \\ a_{m1} & a_{m2} & \dots & a_{mn} \end{bmatrix},$$

or in short (a_{ij}).

If a matrix contains only one row, it is called a **row matrix**. A matrix with one column is called **column matrix**. If the number of rows equals the number of columns of a matrix A, the matrix is called a **square matrix**.

For example, the following are matrices:

$$A = \begin{bmatrix} 2 & 0 & -4 \\ 3 & -5 & 6 \end{bmatrix}.$$

$$B = \begin{bmatrix} 3.1 & 2.6 \\ 0 & \pi \\ \sqrt{2} & \frac{1}{2} \end{bmatrix}.$$

$$C = \begin{bmatrix} 2 & 4 & -6 \\ 1 & 0 & 3 \\ 9 & -2 & 5 \end{bmatrix}.$$

$$D = \begin{bmatrix} 6 & 2 & -9 & 4 \end{bmatrix}.$$

$$E = \begin{bmatrix} -4 \\ 0 \\ 3 \end{bmatrix}.$$

Matrix A is 2×3, B is 4×2, C is 3×3, D is 1×4, and E is 3×1. For example, $a_{13} = -4$, $a_{21} = 3$, $b_{22} = \pi$, $b_{31} = \sqrt{2}$, $c_{32} = -2$, $d_{13} = -9$, and $e_{31} = 3$. Matrix C is a square matrix of size 3×3.

The **diagonal entries** of a square matrix of size $n \times n$ are $a_{11}, a_{22}, a_{33}, \ldots, a_{nn}$. The diagonal entries of the matrix C are 2, 0, 5.

3.3.2. Matrix Equality

Two matrices A and B are equal if and only if they have the same size and their corresponding elements are equal, that is, $a_{ij} = b_{ij}$ for each i and j. For example, let

$$A = \begin{bmatrix} 2 & 4 \\ 6 & 9 \end{bmatrix}.$$

$$B = \begin{bmatrix} 2 & 4 \\ 6 & 9 \end{bmatrix}.$$

$$C = \begin{bmatrix} a & 4 \\ 6 & b \end{bmatrix}.$$

$$D = \begin{bmatrix} 2 & 4 & 0 \\ 6 & 9 & 0 \end{bmatrix}.$$

Then $A=B$, and $A=C$ if $a=2$ and $b=9$. $A \neq D$ and $B \neq D$ because A and B are 2×2 matrices and D is 2×3.

3.3.3. Scalar Multiplication

If A is an $m\times n$ matrix and α is a scalar, then the scalar multiple of A by α is an $m\times n$ matrix obtained by multiplying each element of A by α. That is $\alpha A=(\alpha a_{ij})$. For example if $A=\begin{bmatrix} 2 & -6 & 8 \\ -4 & 4 & 12 \end{bmatrix}$, then $\frac{1}{2}A=\begin{bmatrix} 1 & -3 & 4 \\ -2 & 2 & 6 \end{bmatrix}$ and $2A=\begin{bmatrix} 4 & -12 & 16 \\ -8 & 8 & 24 \end{bmatrix}$.
Matrix $(-1)A$, written as $-A$, is called the negation of A.

3.3.4. Matrix Addition

Let A and B be two matrices with the same dimension, say $m\times n$. The **sum of** A **and** B, written $A+B$, is the $m\times n$ matrix in which each (i, j) entry is the sum of the corresponding entries of A and B—that is $a_{ij}+b_{ij}$ for each i and j. The **difference** of A and B, written $A-B$, is defined as $A+(-1)B$.

For example, let $A=\begin{bmatrix} 4 & -1 \\ 6 & 5 \\ -7 & 2 \end{bmatrix}$, $B=\begin{bmatrix} 2 & 3 \\ -4 & 1 \\ 9 & 2 \end{bmatrix}$, and $C=\begin{bmatrix} 1 & 2 & 3 \\ 4 & 5 & 6 \\ 7 & 8 & 9 \end{bmatrix}$. Then

$A+B=\begin{bmatrix} 4+2 & -1+3 \\ 6-4 & 5+1 \\ -7+9 & 2+2 \end{bmatrix}=\begin{bmatrix} 6 & 2 \\ 2 & 6 \\ 2 & 4 \end{bmatrix}$, whereas $A+C$ is not defined because A and C have different sizes. Similarly $B+C$ is not defined.

$$A-B=\begin{bmatrix} 4-2 & -1-3 \\ 6-(-4) & 5-1 \\ -7-9 & 2-2 \end{bmatrix}=\begin{bmatrix} 2 & -4 \\ 2 & 4 \\ -11 & 0 \end{bmatrix}$$

and

$$B-A=\begin{bmatrix} 2-4 & 3-(-1) \\ -4-6 & 1-5 \\ -9-(-7) & 2-2 \end{bmatrix}=\begin{bmatrix} -2 & 4 \\ -10 & -4 \\ -2 & 0 \end{bmatrix}.$$

Example 3.9
Find values for t, x, y, and z that satisfy the following matrix equation:

$$\begin{bmatrix} 2 & -1 \\ y & 3 \\ 9 & 7 \end{bmatrix} + \begin{bmatrix} x & 6 \\ -4 & 5 \\ -t & -2 \end{bmatrix} = \begin{bmatrix} 6 & 5 \\ 2 & 8 \\ 1 & z \end{bmatrix}.$$

Solution
We have

$$\begin{bmatrix} 2+x & -1+6 \\ y-4 & 3+5 \\ 9-t & 7-2 \end{bmatrix} = \begin{bmatrix} 6 & 5 \\ 2 & 8 \\ 1 & z \end{bmatrix}.$$

These two matrices have the same size and they are equal if and only if the corresponding elements are the same. Therefore,

$$2+x=6.$$
$$y-4=2.$$
$$9-t=1.$$
$$7-2=z.$$

These imply that:

$$x=6-2=4.$$
$$y=2+4=6.$$
$$t=9-1=8.$$
$$z=5.$$

3.3.5. Matrix Multiplication

Matrix addition and subtraction were defined by adding or subtracting the corresponding elements of two matrices of the same size. It might look natural to define the product of two matrices of the same size by multiplying the corresponding elements. However, this product has very few applications. Before we give the definition of the product of two matrices we need to know how to multiply a row matrix with a column matrix of the same dimension.

Let $R=\begin{bmatrix} r_1 & r_2 & \cdots & r_n \end{bmatrix}$ be a row matrix of dimension $1 \times n$ and $C=\begin{bmatrix} c_1 \\ c_2 \\ \vdots \\ c_n \end{bmatrix}$ be a column matrix of dimension $n \times 1$. Then the product RC is defined as

$$RC = \begin{bmatrix} r_1 & r_2 & \dots & r_n \end{bmatrix} \begin{bmatrix} c_1 \\ c_2 \\ \vdots \\ c_n \end{bmatrix} = r_1c_1 + r_2c_2 + \dots + r_nc_n.$$

For example,

$$[2 \quad -3 \quad 4] \begin{bmatrix} 5 \\ 1 \\ -2 \end{bmatrix} = 2 \cdot 5 + (-3) \cdot 1 + 4 \cdot (-2) = -1.$$

$$[2 \quad 1 \quad 0 \quad 5] \begin{bmatrix} \frac{1}{2} \\ -2 \\ 4 \\ 1 \end{bmatrix} = 2 \cdot \frac{1}{2} + 1 \cdot (-2) + 0 \cdot 4 + 5 \cdot 1 = 4.$$

$[2 \quad -3 \quad 4] \begin{bmatrix} 5 \\ 3 \end{bmatrix}$ is undefined because the dimension of the row matrix does not equal the dimension of the column matrix.

Now we give the definition of the product of two matrices.

Definition

Let A be an $m \times r$ matrix and B be an $r \times n$ matrix. The **product** AB is defined to be the $m \times n$ matrix C, where c_{ij} equals the product of the ith row of A and the jth column of B. The size of C is $m \times n$.

Note that the product AB is defined only if the number of columns of A is the same as the number of rows of B.

Example 3.10

Find the product AB if $A = \begin{bmatrix} 3 & 1 & -2 \\ 4 & 0 & 2 \end{bmatrix}$ and $B = \begin{bmatrix} 2 & 5 & 1 & 6 \\ 1 & -4 & 0 & 4 \\ -3 & 1 & -4 & 2 \end{bmatrix}$.

Solution

Before we perform the product we need to check that the product AB is defined. Because A is 2×3 and B is 3×4, the product AB is defined and is a 2×4 matrix. Let $AB = C$. The elements of C are defined as noted earlier. For example, c_{12} = row 1 of A times column 2 of $B=[3\quad 1\quad -2]\begin{bmatrix}5\\-4\\1\end{bmatrix}=3\cdot 5+1\cdot(-4)+(-2)\cdot 1=2$, and

$c_{24}=[4\quad 0\quad 2]\begin{bmatrix}6\\4\\2\end{bmatrix}=4\cdot 6+0\cdot 4+2\cdot 2=28.$

We have

$$AB=C=\begin{bmatrix}3 & 1 & -2\\4 & 0 & 2\end{bmatrix}\begin{bmatrix}2 & 5 & 1 & 6\\1 & -4 & 0 & 4\\-3 & 1 & -4 & 2\end{bmatrix}=$$

$$=\begin{bmatrix}3\cdot 2+1\cdot 1+(-2)\cdot(-3) & 3\cdot 5+1\cdot(-4)+(-2)\cdot 1 & 3\cdot 1+1\cdot 0+(-2)\cdot(-4) & 3\cdot 6+1\cdot 4+(-2)\cdot 2\\4\cdot 2+0\cdot 1+2\cdot(-3) & 4\cdot 5+0\cdot(-4)+2\cdot 1 & 4\cdot 1+0\cdot 0+2\cdot(-4) & 4\cdot 6+0\cdot 4+2\cdot 2\end{bmatrix}$$

$$=\begin{bmatrix}13 & 9 & 11 & 18\\2 & 22 & -4 & 28\end{bmatrix}.$$

3.3.6. Special Matrices

A **zero matrix** of size $m\times n$ is a matrix with m rows and n columns, all of whose elements are zero. We use the notation O_{mn} to represent the $m\times n$ zero matrix. For example, the zero matrix of size 2×4 is $\begin{bmatrix}0 & 0 & 0 & 0\\0 & 0 & 0 & 0\end{bmatrix}$.

A **diagonal matrix** is a square matrix in which all the elements not in the diagonal are zeros. For example $\begin{bmatrix}2 & 0 & 0\\0 & 4 & 0\\0 & 0 & -3\end{bmatrix}$ is a 3×3 diagonal matrix.

An **identity matrix** is a diagonal matrix in which the entries in the diagonal are ones. We use the notation I_n to represent the $n\times n$ identity matrix. For example,

$$I_2=\begin{bmatrix}1 & 0\\0 & 1\end{bmatrix}.$$

$$I_4=\begin{bmatrix}1 & 0 & 0 & 0\\0 & 1 & 0 & 0\\0 & 0 & 1 & 0\\0 & 0 & 0 & 1\end{bmatrix}.$$

Note that if α is a scalar, then $\alpha \cdot 1 = 1 \cdot \alpha = \alpha$. The identity matrix plays the role of 1 for real numbers. If A is an $n \times n$ square matrix, then

$$AI_n = I_n A = A.$$
$$AO_n = O_n A = O_n.$$

The zero matrix plays the role of 0 for real numbers. Let A be an $m \times n$ matrix and O be the $m \times n$ zero matrix; then

$$A + O = O + A = A.$$
$$A - A = O.$$

The following properties can be verified: If A, B, and C are three matrices of the same size, then

1. $A + B = B + A$.
2. $A + (B + C) = (A + B) + C$.

3.3.7. Systems of Linear Equations

We will use the product of matrices to represent a system of linear equations in matrix notation. Consider the following system of equations.

$$2x_1 + x_2 + 6x_3 = -9.$$
$$-3x_1 + x_2 + 5x_3 = -9.$$
$$5x_1 + 8x_2 + x_3 = 5.$$

Represent each side as a column matrix:

$$\begin{bmatrix} 2x_1 + 4x_2 + 6x_3 \\ -3x_1 + x_2 + 5x_3 \\ 5x_1 + 8x_2 + x_3 \end{bmatrix} = \begin{bmatrix} 6 \\ -9 \\ 5 \end{bmatrix}.$$

Because the left matrix can be represented by the product of two matrices, we have

$$\begin{bmatrix} 2 & 4 & 6 \\ -3 & 1 & 5 \\ 5 & 8 & 1 \end{bmatrix} \begin{bmatrix} x_1 \\ x_2 \\ x_3 \end{bmatrix} = \begin{bmatrix} 6 \\ -9 \\ 5 \end{bmatrix}.$$

Letting

$$A = \begin{bmatrix} 2 & 4 & 6 \\ -3 & 1 & 5 \\ 5 & 8 & 1 \end{bmatrix}.$$

$$X = \begin{bmatrix} x_1 \\ x_2 \\ x_3 \end{bmatrix}.$$

$$B = \begin{bmatrix} 6 \\ -9 \\ 5 \end{bmatrix}.$$

This system of linear equations is represented by the matrix equation $AX=B$. Matrix A is called the **coefficient matrix**.

In general, a system of m linear equations in n variables,

$$\begin{array}{cccc} a_{11}x_1 + a_{12}x_2 + \cdots + a_{1n}x_n = b_1 \\ a_{21}x_1 + a_{22}x_2 + \cdots + a_{2n}x_n = b_2 \\ \vdots \quad \vdots \quad \vdots \quad \vdots \\ a_{m1}x_1 + a_{m2}x_2 + \cdots + a_{mn}x_n = b_m \end{array}$$

can be represented in the matrix equation $AX=B$, where

$$A = \begin{bmatrix} a_{11} & a_{12} & \cdots & a_{1n} \\ a_{21} & a_{22} & \cdots & a_{2n} \\ \vdots & \vdots & \cdots & \vdots \\ a_{m1} & a_{m2} & \cdots & a_{mn} \end{bmatrix}.$$

$$X = \begin{bmatrix} x_1 \\ x_2 \\ \vdots \\ x_n \end{bmatrix}.$$

$$B = \begin{bmatrix} b_1 \\ b_2 \\ \vdots \\ b_m \end{bmatrix}.$$

3.3.8. Matrix Powers

If A is an $n \times n$ matrix, then $A^2 = AA$. In general $A^k = \underbrace{AA \cdots A}_{k \text{ times}}$. That is, for any positive integer k, A^k is A multiplied by itself k times. We have $A^1 = A$, and will define $A^0 = I_n$. From this definition the following properties can be verified.

$$A^n A^m = A^{n+m}.$$

$$(A^n)^m = A^{nm}.$$

For example if $A = \begin{bmatrix} 2 & 1 \\ 0 & 3 \end{bmatrix}$, then $A^2 = \begin{bmatrix} 2 & 1 \\ 0 & 3 \end{bmatrix}\begin{bmatrix} 2 & 1 \\ 0 & 3 \end{bmatrix} = \begin{bmatrix} 4 & 5 \\ 0 & 9 \end{bmatrix}$, and $A^3 = A^2 A =$ $\begin{bmatrix} 4 & 5 \\ 0 & 9 \end{bmatrix}\begin{bmatrix} 2 & 1 \\ 0 & 3 \end{bmatrix} = \begin{bmatrix} 8 & 19 \\ 0 & 27 \end{bmatrix}$.

3.3.9. Matrix Transpose

If A is an $m \times n$ matrix, then the **transpose** of A, denoted by A^t (or A') is an $n \times m$ matrix, in which the rows of A are the columns of A^t. That is, the ith row of A is the ith column of A^t.

For example, let

$$A = \begin{bmatrix} 4 & 0 & -3 \\ 2 & -6 & 1 \end{bmatrix}.$$

$$B = \begin{bmatrix} 2 & 1 & 4 & 1.25 \\ 0 & -3 & 0.5 & 9 \\ 5 & 6 & 2 & 4 \end{bmatrix}.$$

$$C = [1 \quad 3 \quad 5].$$

$$D = \begin{bmatrix} 2 \\ 0 \\ 1 \\ 7 \end{bmatrix}.$$

Then

$$A^t = \begin{bmatrix} 4 & 2 \\ 0 & -6 \\ -3 & 1 \end{bmatrix}.$$

$$B^t = \begin{bmatrix} 2 & 0 & 5 \\ 1 & -3 & 6 \\ 4 & 0.5 & 2 \\ 1.25 & 9 & 4 \end{bmatrix}.$$

$$C^t = \begin{bmatrix} 1 \\ 3 \\ 5 \end{bmatrix}.$$

$$D^t = [2 \quad 0 \quad 1 \quad 7].$$

Note the transpose of thc matrix A in MATLAB is A′.

3.3.10. Matrix Operations with MATLAB

Matrix Operation	Operation in MATLAB	Example
Addition, $A+B$	A+B	A+B
Subtraction, $A-B$	A−B	A−B
Multiplication, AB	A*B	A*B
Scalar multiplication, cA	c*A	5*A
Matrix power, A^k	A^k	A^4
Matrix transpose A'	A′	A′

Special Matrices

Matrix Operation	Operation in MATLAB	Example
Identity matrix, I_n	eye(n)	eye(4)
$m \times n$ zero matrix, O_{mn}	zeros(m, n)	zeros(2, 4)
$m \times n$ matrix all ones	ones(m, n)	ones(3, 4)
$m \times n$ random matrix	rand(m, n)	rand(2, 3)

The random elements are between 0 and 1.

Row i of matrix A	A(i, :)	A(3, :)
Column j of matrix A	A(:, j)	A(:, 2)

Here are MATLAB commands to illustrate some of the matrix terminology and operations:

```
>> A=[4 -1;6 5; -7 2]                % Enter the 3×2 matrix A
```

```
A =
    4  -1
    6   5
   -7   2

>> B=[2 3; -4 1;9 2]            % Enter the 3×2 matrix B
B =
    2 3
   -4 1
    9 2

>> C=[1 2 3; 4 5 6; 7 8 9]      % Enter the 3×3 matrix C
C =
   1 2 3
   4 5 6
   7 8 9

>> A+B                          % The sum A+B
ans =
   6 2
   2 6
   2 4

>> B+A                          % The sum B+A
ans =
   6 2
   2 6
   2 4

>> (A+B)-(B+A)                  % To test if A+B=B+A
ans =
   0 0
   0 0
   0 0

>> A+B==B+A                     % Another way to test if A+B=B+A
ans =                           % MATLAB compares the corresponding
   1 1                          % components of both sides and returns
   1 1                          % 1 if they are equal otherwise it returns 0
   1 1

>> A-B                          % The difference A-B
ans =
     2  -4
    10   4
   -16   0
```

```
>> B-A                                  % The difference B-A
ans =
   -2   4
  -10  -4
   16   0

>> A+C                                  % The sum A+C is not defined
Error using +                           % Note that the error messages will
Matrix dimensions must agree.           % appear on the computer screen in red

>> (A-B)==(B-A)                         % To test if A - B = B - A
ans =
  0 0
  0 0
  0 1

>> (A-B)-(B-A)                          % Another way to test if A - B = B - A
ans =
    4  -8
   20   8
  -32   0

>> 2*A                                  % The scalar product 2A
ans =
    8  -2
   12  10
  -14   4

>> 1/2*B                                % The scalar product 1/2B
ans =
   1.0000 1.5000
  -2.0000 0.5000
   4.5000 1.0000

>> [2 -3 4]*[5; 1; -2]                  % The product of a 1 x 3 row and a 3 x 1
>>                                      % column
ans =
  -1

>> [2 1 0 5]*[1/2; -2; 4; 1]            % The product of a 1 x 4 row and a 4 x 1
>>                                      % column
ans =
  4

>> [2 -3 4]*[5;3]                       % The product of 1 x 3 row and 2 x 1 column
Error using *
Inner matrix dimensions must agree.
```

```
>> A=[3 1 -2; 4 0 2]                          % Enter 2 x 3 matrix A
A=
   3 1 -2
   4 0  2

>> B=[2 5 1 6; 1 -4 0 4; -3 1 -4 2]           % Enter 3 x 3 matrix B
B=
    2  5  1 6
    1 -4  0 4
   -3  1 -4 2

>> A*B                                        % The product AB
ans=
   13  9 11 18
    6 22 -4 28

>> C=A*B                                      % C=AB
C=
   13  9 11 18
    6 22 -4 28

>> zeros(2, 4)                                % 2 x 4 zero matrix
ans=
   0 0 0 0
   0 0 0 0

>> zeros(2)                                   % 2 x 2 zero matrix
ans=
   0 0
   0 0

>> eye(4)                                     % The identity matrix I4
ans=
   1 0 0 0
   0 1 0 0
   0 0 1 0
   0 0 0 1

>> A=[2 1; 0 3]
A=
   2 1
   0 3

>> A^2
ans=
   4 5
   0 9
```

```
>> A^3
ans =
  8 19
  0 27

>> A^10
ans =
  1024 58025
     0 59049

>> ones(3,4)
ans =
  1 1 1 1
  1 1 1 1
  1 1 1 1

>> rand(2,3)
ans =
  0.8147 0.1270 0.6324
  0.9058 0.9134 0.0975


>> A=[4 0 -3; 2 -6 1];
>> B=[2 1 4 1.25; 0 -3 0.5 9; 5 6 2 4];
>> C=[1 3 5];
>> D=[2; 0; 1; 7];
>> A'
ans =
   4  2
   0 -6
  -3  1
>> B'
ans =
       2   0 5
       1  -3 6
       4 0.5 2
    1.25   9 4
>> C'
ans =
  1
  3
  5
>> D'
ans =
  2 0 1 7
```

3.3.11. Model 3.6: Population Movement, Part I

We consider a simple model of population movement between a certain city and its surrounding suburbs. For simplicity we assume the following:

1. The people who move from the city go to the suburbs and the people who move from the suburbs go to the city.
2. During a year, the total population in the city and its surrounding suburb is fixed—that is we ignore other factors such as births and deaths.

Under these two assumptions, the total population in the city and the suburb is the same every year. Assume that that the demographic studies showed that during 2010, 6% of the city population moved to suburb and 2% of the suburb population moved to the city. This means that 94% of city population stayed in the city and 98% of the suburb population stayed in the suburb. Assume that this trend continues and the migration percentages remain constant. Let's consider 2010 to be the initial year and assume that the initial populations (at 2010) in the city and suburbs are 3 million and 7 million, respectively.

Discussion

Let C_n and S_n be the city and the suburb populations after n years from 2010, respectively. The populations can be modeled by the following difference equations:

$$C_n = 0.94C_{n-1} + 0.02S_{n-1} \tag{3.1}$$

$$S_n = 0.06C_{n-1} + 0.98S_{n-1} \tag{3.2}$$

This system can be represented by the following matrix difference equation:

$$X_n = TX_{n-1}, \tag{3.3}$$

where $X_n = \begin{bmatrix} C_n \\ S_n \end{bmatrix}$ and $T = \begin{bmatrix} 0.94 & 0.02 \\ 0.06 & 0.98 \end{bmatrix}$. Note that $X_0 = \begin{bmatrix} C_0 \\ S_0 \end{bmatrix} = \begin{bmatrix} 3 \\ 7 \end{bmatrix}$ and the unit is in millions.

Equation 3.3 is a first-order linear homogenous difference equation. Column matrix X_n is called the **distribution vector**, and matrix T is called the **transition matrix**.

The solution X_n, $n=1, 2, 3, \ldots$ may be obtained in two different ways, either by iterating equation 3.3 or by using the analytical solution of equation 3.3. For example to find X_3 by iteration we have

$$X_1 = TX_0 = \begin{bmatrix} 0.94 & 0.02 \\ 0.06 & 0.98 \end{bmatrix}\begin{bmatrix} 3 \\ 7 \end{bmatrix} = \begin{bmatrix} 0.94 \cdot 3 + 0.02 \cdot 7 \\ 0.06 \cdot 3 + 0.98 \cdot 7 \end{bmatrix} = \begin{bmatrix} 2.96 \\ 7.04 \end{bmatrix}.$$

$$X_2 = TX_1 = \begin{bmatrix} 0.94 & 0.02 \\ 0.06 & 0.98 \end{bmatrix}\begin{bmatrix} 2.96 \\ 7.04 \end{bmatrix} = \begin{bmatrix} 2.9232 \\ 7.0768 \end{bmatrix}.$$

$$X_3 = TX_2 = \begin{bmatrix} 0.94 & 0.02 \\ 0.06 & 0.98 \end{bmatrix} \begin{bmatrix} 2.9232 \\ 7.0768 \end{bmatrix} = \begin{bmatrix} 2.8893 \\ 7.1107 \end{bmatrix}.$$

The analytical solution of the equation 3.3 is

$$X_n = T^n X_0.$$

For example to find X_3 using the analytical solution, we have

$$X_3 = T^3 X_0 = \begin{bmatrix} 0.94 & 0.02 \\ 0.06 & 0.08 \end{bmatrix}^3 \begin{bmatrix} 3 \\ 7 \end{bmatrix} = \begin{bmatrix} 0.8340 & 0.0553 \\ 0.1660 & 0.9447 \end{bmatrix} \begin{bmatrix} 3 \\ 7 \end{bmatrix} = \begin{bmatrix} 2.8893 \\ 7.1107 \end{bmatrix}.$$

3.3.12. Inverse of a Square Matrix

In this section, I introduce the concept of the inverse of a square matrix. The inverse of a matrix is a powerful tool for solving a system of linear equations, when the number of variables is the same as the number of equations.

Consider the following scalar equation:

$$ab = 1,$$

where a and b are real numbers, and b is called the **multiplicative inverse** of a. We write

$$b = a^{-1},$$

and a is the multiplicative inverse of b:

$$a = b^{-1}.$$

For example, 6^{-1} is the multiplicative inverse of 6 because

$$6(6^{-1}) = 6\left(\frac{1}{6}\right) = 1.$$

We extend the multiplicative inverse of real numbers to the inverse of a square matrix.

Definition

Let A be a square matrix of size $n \times n$. If there exists an $n \times n$ matrix B such that

$$AB = BA = I_n,$$

then B is called the **inverse** of A, and we write

$$A^{-1} = B.$$

If A^{-1} (read "A inverse") exists, the matrix A is said to be **invertible**. Note that A^{-1} cannot be written as $\dfrac{1}{A}$.

Example 3.11
Let

$$A = \begin{bmatrix} 4 & 7 \\ 1 & 2 \end{bmatrix}.$$

Show that

$$B = \begin{bmatrix} 2 & -7 \\ -1 & 4 \end{bmatrix}$$

is the inverse of A.

Solution
To show that B is the inverse of A, we need to show that

$$AB = BA = I_2.$$

We have

$$\begin{aligned} AB &= \begin{bmatrix} 4 & 7 \\ 1 & 2 \end{bmatrix}\begin{bmatrix} 2 & -7 \\ -1 & 4 \end{bmatrix} \\ &= \begin{bmatrix} 4 \cdot 2 + 7(-1) & 4(-7) + 7 \cdot 4 \\ 1 \cdot 2 - 2(-1) & 1(-7) + 2 \cdot 4 \end{bmatrix} \\ &= \begin{bmatrix} 1 & 0 \\ 0 & 1 \end{bmatrix} = I_2 \end{aligned}$$

and

$$BA = \begin{bmatrix} 2 & -7 \\ -1 & 4 \end{bmatrix}\begin{bmatrix} 4 & 7 \\ 1 & 2 \end{bmatrix} = \begin{bmatrix} 1 & 0 \\ 0 & 1 \end{bmatrix} = I_2.$$

Therefore $A^{-1} = B$. Note that not every square matrix has an inverse.

Example 3.12
Show that the matrix

$$A = \begin{bmatrix} 1 & 2 \\ 3 & 6 \end{bmatrix}$$

does not have an inverse.

Solution

If A has an inverse, say it is B,

$$B = \begin{bmatrix} a & c \\ b & d \end{bmatrix},$$

where a, b, c, and d are real numbers, then

$$AB = I_2.$$

We will show that this is not true. Assuming that $AB = I_2$, we have

$$\begin{bmatrix} 1 & 2 \\ 3 & 6 \end{bmatrix} \begin{bmatrix} a & c \\ b & d \end{bmatrix} = \begin{bmatrix} 1 & 0 \\ 0 & 1 \end{bmatrix}.$$

$$\begin{bmatrix} a+2b & c+2d \\ 3a+6b & 3c+6d \end{bmatrix} = \begin{bmatrix} 1 & 0 \\ 0 & 1 \end{bmatrix}.$$

Equating the elements in the first column, we get

$$a + 2b = 1.$$
$$3a + 6b = 0.$$

Because these two equations represent two parallel lines, there is no solution for this system. Similarly, equating the elements in the second column we get the system

$$c + 2d = 0.$$
$$3c + 6d = 1.$$

This system also has no solution. Consequently, there are no real numbers a, b, c, and d, such that

$$AB = \begin{bmatrix} 1 & 2 \\ 3 & 6 \end{bmatrix} \begin{bmatrix} a & c \\ b & d \end{bmatrix} = \begin{bmatrix} 1 & 0 \\ 0 & 1 \end{bmatrix}.$$

Therefore, A has no inverse.

3.3.13. Finding a Matrix Inverse

We will investigate a method to find the inverse of a matrix if it is invertible. If the inverse of a matrix does not exist, the method will indicate so.

Consider

$$A = \begin{bmatrix} 4 & 7 \\ 1 & 2 \end{bmatrix}.$$

The inverse A^{-1} will be a 2×2 matrix. Let's assume that $A^{-1}=B$,

$$B=\begin{bmatrix} a & c \\ b & d \end{bmatrix},$$

where a, b, c, and d are unknown real numbers. Because B is the inverse of A, we have

$$AB=I_2.$$

$$\begin{bmatrix} 4 & 7 \\ 1 & 2 \end{bmatrix}\begin{bmatrix} a & c \\ b & d \end{bmatrix}=\begin{bmatrix} 1 & 0 \\ 0 & 1 \end{bmatrix}.$$

$$\begin{bmatrix} 4a+7b & 4c+7d \\ a+2b & c+2d \end{bmatrix}=\begin{bmatrix} 1 & 0 \\ 0 & 1 \end{bmatrix}.$$

Equating the corresponding elements of these two matrices, we get

$$4a+7b=1 \qquad 4c+7d=0.$$

$$a+2b=0 \qquad c+2d=1.$$

We have two systems. One system with two variables, a and b, and coefficient matrix

$$\begin{bmatrix} 4 & 7 \\ 1 & 2 \end{bmatrix}.$$

To find a and b rows reduce the augmented matrix

$$\begin{bmatrix} 4 & 7 & 1 \\ 1 & 2 & 0 \end{bmatrix}.$$

The other system has variables c and d. The coefficient matrix is

$$\begin{bmatrix} 4 & 7 \\ 1 & 2 \end{bmatrix}.$$

Similarly, to find c and d you can row reduce the augmented matrix,

$$\begin{bmatrix} 4 & 7 & 0 \\ 1 & 2 & 1 \end{bmatrix}.$$

Because the two systems have the same coefficient matrix, it is more efficient to solve the two systems by obtaining the reduced echelon form of the augmented matrix,

$$\left[\begin{array}{cc|cc} 4 & 7 & 1 & 0 \\ 1 & 2 & 0 & 1 \end{array}\right].$$

Let's row reduce this augmented matrix:

$$\left[\begin{array}{cc|cc} 4 & 7 & 1 & 0 \\ 1 & 2 & 0 & 1 \end{array}\right] \qquad R_1 \leftrightarrow R_2$$

$$\approx \left[\begin{array}{cc|cc} 1 & 2 & 0 & 1 \\ 4 & 7 & 1 & 0 \end{array}\right] \qquad R_2 \leftarrow R_2 + (-4)R_1$$

$$\approx \left[\begin{array}{cc|cc} 1 & 2 & 0 & 1 \\ 0 & -1 & 1 & -4 \end{array}\right] \qquad R_2 \leftarrow (-1)R_1$$

$$\approx \left[\begin{array}{cc|cc} 1 & 2 & 0 & 1 \\ 0 & 1 & -1 & 4 \end{array}\right] \qquad R_1 \leftarrow R_1 + (-2)R_2$$

$$\approx \left[\begin{array}{cc|cc} 1 & 0 & 2 & -7 \\ 0 & 1 & -1 & 4 \end{array}\right].$$

Therefore, the solutions of the two systems are

$$a = 2 \qquad c = -7.$$

$$b = -1 \qquad d = 4.$$

Consequently, the inverse matrix A^{-1} is

$$A^{-1} = \begin{bmatrix} 2 & -7 \\ -1 & 4 \end{bmatrix}.$$

Note that in Example 3.11 you verified this result. This procedure can be generalized to find the inverse of a square matrix of any size.

3.3.14. Finding the Inverse of a Square Matrix

Let A be an $n \times n$ matrix. To find A^{-1}:

1. Form the augmented matrix $[A \mid I_n]$, where I_n is the $n \times n$ identity matrix.
2. Use row operations to obtain the reduced echelon form of the augmented matrix $[A \mid I_n]$.
3. If the reduced echelon form is in the form $[I_n \mid B]$, then $B = A^{-1}$.
4. If the reduced echelon form is not in the form $[I_n \mid B]$, then A has no inverse.

Example 3.13

Find the inverse, if it exists, of the matrix

$$A = \begin{bmatrix} -1 & 3 & 3 \\ 1 & -2 & -2 \\ 2 & -5 & -4 \end{bmatrix}.$$

Solution

Form the augmented matrix $[A \mid I_3]$ and use row operations to obtain the reduced echelon form. We have

$$\left[\begin{array}{ccc|ccc} -1 & 3 & 3 & 1 & 0 & 0 \\ 1 & -2 & -2 & 0 & 1 & 0 \\ 2 & -5 & -4 & 0 & 0 & 1 \end{array}\right] \qquad R_1 \leftarrow (-1)R_1$$

$$\approx \left[\begin{array}{ccc|ccc} 1 & -3 & -3 & -1 & 0 & 0 \\ 1 & -2 & -2 & 0 & 1 & 0 \\ 2 & -5 & -4 & 0 & 0 & 1 \end{array}\right] \qquad \begin{array}{l} R_2 \leftarrow R_2 + (-1)R_1 \\ R_3 \leftarrow R_3 + (-2)R_1 \end{array}$$

$$\approx \left[\begin{array}{ccc|ccc} 1 & -3 & -3 & -1 & 0 & 0 \\ 0 & 1 & 1 & 1 & 1 & 0 \\ 0 & 1 & 2 & 2 & 0 & 1 \end{array}\right] \qquad \begin{array}{l} R_1 \leftarrow R_1 + (3)R_2 \\ \\ R_3 \leftarrow R_3 + (-1)R_2 \end{array}$$

$$\approx \left[\begin{array}{ccc|ccc} 1 & 0 & 0 & 2 & 3 & 0 \\ 0 & 1 & 1 & 1 & 1 & 0 \\ 0 & 0 & 1 & 1 & -1 & 1 \end{array}\right] \qquad R_2 \leftarrow R_2 + (-1)R_3$$

$$\approx \left[\begin{array}{ccc|ccc} 1 & 0 & 0 & 2 & 3 & 0 \\ 0 & 1 & 0 & 0 & 2 & -1 \\ 0 & 0 & 1 & 1 & -1 & 1 \end{array}\right].$$

Thus

$$A^{-1} = \begin{bmatrix} 2 & 3 & 0 \\ 0 & 2 & -1 \\ 1 & -1 & 1 \end{bmatrix}.$$

To check that this is true, we have

$$AA^{-1} = \begin{bmatrix} -1 & 3 & 3 \\ 1 & -2 & -2 \\ 2 & -5 & -4 \end{bmatrix} \begin{bmatrix} 2 & 3 & 0 \\ 0 & 2 & -1 \\ 1 & -1 & 1 \end{bmatrix} = \begin{bmatrix} 1 & 0 & 0 \\ 0 & 1 & 0 \\ 0 & 0 & 1 \end{bmatrix} = I_3$$

and

$$A^{-1}A = \begin{bmatrix} 2 & 3 & 0 \\ 0 & 2 & -1 \\ 1 & -1 & 1 \end{bmatrix} \begin{bmatrix} -1 & 3 & 3 \\ 1 & -2 & -2 \\ 2 & -5 & -4 \end{bmatrix} = \begin{bmatrix} 1 & 0 & 0 \\ 0 & 1 & 0 \\ 0 & 0 & 1 \end{bmatrix} = I_3.$$

Example 3.14

Find the inverse, if it exists, of the matrix

$$A = \begin{bmatrix} 1 & 1 & -1 \\ -2 & -1 & 4 \\ 3 & 4 & -1 \end{bmatrix}.$$

Solution

Form the augmented matrix $[A \mid I_3]$ and use row operations to obtain the reduced echelon form $[I_3 \mid B]$ if possible. We have

$$\left[\begin{array}{ccc|ccc} 1 & 1 & -1 & 1 & 0 & 0 \\ -2 & -1 & 4 & 0 & 1 & 0 \\ 3 & 4 & -1 & 0 & 0 & 1 \end{array}\right] \qquad \begin{array}{l} R_2 \leftarrow R_2 + (2)R_1 \\ R_3 \leftarrow R_3 + (-3)R_1 \end{array}$$

$$\approx \left[\begin{array}{ccc|ccc} 1 & 1 & -1 & 1 & 0 & 0 \\ 0 & 1 & 2 & 2 & 1 & 0 \\ 0 & 1 & 2 & -3 & 0 & 1 \end{array}\right] \qquad \begin{array}{l} R_1 \leftarrow R_1 + (-1)R_2 \\ \\ R_3 \leftarrow R_3 + (-1)R_2 \end{array}$$

$$\approx \left[\begin{array}{ccc|ccc} 1 & 0 & -3 & -1 & -1 & 0 \\ 0 & 1 & 2 & 2 & 1 & 0 \\ 0 & 0 & 0 & -5 & -1 & 1 \end{array}\right].$$

The last row indicates that we cannot get 1 in the position (3, 3) to reduce the augmented matrix into $[I_3 \mid B]$. Thus A^{-1} does not exist.

3.3.15. Inverse of a Square Matrix in MATLAB

Using MATLAB, the inverse of a square matrix A may be calculated by the method just described or by using the MATLAB function **inv(A)**. For example to find the inverse of matrix $A = \begin{bmatrix} -1 & 3 & 3 \\ 1 & -2 & -2 \\ 2 & -5 & -4 \end{bmatrix}$.

```
>> A=[-1 3 3; 1 -2 -2; 2 -5 -4]   % Enter the matrix A
A=
  -1  3  3
   1 -2 -2
   2 -5 -4
>> rref([A eye(3)])               % Find the row reduced echelon form
ans=                              % of the augmented matrix [A|I3]
  1 0 0 2  3  0
  0 1 0 0  2 -1
  0 0 1 1 -1  1

>> inv(A)                         % Use MATLAB function inv(A) to
ans=                              % find A^-1
  2  3  0
  0  2 -1
  1 -1  1
```

Now find the inverse, if it exists, of matrix $B = \begin{bmatrix} 1 & 1 & -1 \\ -2 & -1 & 4 \\ 3 & 4 & -1 \end{bmatrix}$.

```
>> B=[1 1 -1; -2 -1 4; 3 4 -1]
B=
   1  1 -1
  -2 -1  4
   3  4 -1

>> rref([B eye(3)])
ans=
  1.0000       0 -3.0000       0 -0.8000 -0.2000
       0 1.0000  2.0000       0  0.6000  0.4000
       0      0       0 1.0000  0.2000 -0.2000
>> % This concludes that B^-1 does not exist
>> inv(B)
Warning: Matrix is close to singular or badly scaled.
Results may be inaccurate.
RCOND=3.083953e-18.
```

```
ans =
   1.0e+16*
  -2.7022 -0.5404  0.5404
   1.8014  0.3603 -0.3603
  -0.9007 -0.1801  0.1801
```

3.3.16. Solving a Linear System Using Matrix Inverse

Given a system of linear equations in which the number of equations is the same as the number of unknowns, we will discuss how the inverse matrix can be used to solve the system. We illustrate this method with Example 3.15.

Example 3.15

Use an inverse matrix to solve the following system

$$-x_1 + 3x_2 + 3x_3 = 3.$$
$$x_1 - 2x_2 - 2x_3 = -1.$$
$$2x_1 - 5x_2 - 4x_3 = 4.$$

Solution

This system can be written in the following matrix form

$$\begin{bmatrix} -1 & 3 & 3 \\ 1 & -2 & -2 \\ 2 & -5 & -4 \end{bmatrix} \begin{bmatrix} x_1 \\ x_2 \\ x_3 \end{bmatrix} = \begin{bmatrix} 3 \\ -1 \\ 4 \end{bmatrix}.$$

Setting

$$A = \begin{bmatrix} -1 & 3 & 3 \\ 1 & -2 & -2 \\ 2 & -5 & -4 \end{bmatrix},$$

$$X = \begin{bmatrix} x_1 \\ x_2 \\ x_3 \end{bmatrix},$$

and

$$B = \begin{bmatrix} 3 \\ -1 \\ 4 \end{bmatrix},$$

the system can be written as the matrix equation $AX=B$. If the matrix of coefficients A has an inverse A^{-1}, then multiply both sides of the equation on the left by A^{-1}, to obtain

$$A^{-1}AX = A^{-1}B.$$
$$IX = A^{-1}B.$$
$$X = A^{-1}B.$$

In Example 3.13, we found A^{-1}:

$$A^{-1} = \begin{bmatrix} 2 & 3 & 0 \\ 0 & 2 & -1 \\ 1 & -1 & 1 \end{bmatrix}.$$

We have

$$X = A^{-1}B.$$
$$\begin{bmatrix} x_1 \\ x_2 \\ x_3 \end{bmatrix} = \begin{bmatrix} 2 & 3 & 0 \\ 0 & 2 & -1 \\ 1 & -1 & 1 \end{bmatrix} \begin{bmatrix} 3 \\ -1 \\ 4 \end{bmatrix}.$$
$$\begin{bmatrix} x_1 \\ x_2 \\ x_3 \end{bmatrix} = \begin{bmatrix} 3 \\ -6 \\ 8 \end{bmatrix}.$$

Therefore, the system has the unique solution:

$$x_1 = 3.$$
$$x_2 = -6.$$
$$x_3 = 8.$$

Check this solution.

3.3.17. Model 3.7: Population Movement, Part II

In Model 3.6, the population movement between a city and its surrounding suburb was modeled by the following matrix equation

$$X_n = TX_{n-1},$$

where $T = \begin{bmatrix} 0.94 & 0.02 \\ 0.06 & 0.98 \end{bmatrix}$ is the transition matrix and $X_n = \begin{bmatrix} C_n \\ S_n \end{bmatrix}$ is the population distribution vector, ande C_n and S_n are the city and suburb populations after n years from 2010, respectively. Given the population X_{n-1}, we were able to calculate the next year's population, X_n, by computing the product TX_{n-1}.

Now we have the reverse problem. That is, we know the current year's population, X_n, and we do not know the previous year's population, X_{n-1}. With the inverse matrix, T^{-1}, we will be able to find X_{n-1}. We have

$$X_n = TX_{n-1}.$$

Multiplying both sides of the equation by T^{-1} we get

$$T^{-1}X_n = T^{-1}TX_{n-1} = IX_{n-1} = X_{n-1}.$$

That is, to find X_{n-1}, compute $T^{-1}X_n$. For example, let $X_3 = \begin{bmatrix} 2.8893 \\ 7.1107 \end{bmatrix}$ find X_2. We have

$$T^{-1} = \begin{bmatrix} 1.0652 & -0.0217 \\ -0.0652 & 1.0217 \end{bmatrix}.$$

$$X_2 = T^{-1}X_3 = \begin{bmatrix} 1.0652 & -0.0217 \\ -0.0652 & 1.0217 \end{bmatrix} \begin{bmatrix} 2.8893 \\ 7.1107 \end{bmatrix} = \begin{bmatrix} 2.9232 \\ 7.0768 \end{bmatrix}.$$

Now assume that given X_3 we need to find X_0. We have

$$X_2 = T^{-1}X_3.$$
$$X_1 = T^{-1}X_2 = T^{-1}(T^{-1}X_3) = (T^{-1})^2 X_3.$$
$$X_0 = T^{-1}X_1 = T^{-1}(T^{-1})^2 X_3 = (T^{-1})^3 X_3.$$

Therefore,

$$X_0 = (T^{-1})^3 X_3 = \begin{bmatrix} 1.0652 & -0.0217 \\ -0.0652 & 1.0217 \end{bmatrix}^3 \begin{bmatrix} 2.8893 \\ 7.1107 \end{bmatrix}$$

$$= \begin{bmatrix} 1.2132 & -0.0711 \\ -0.2132 & 1.0711 \end{bmatrix} \begin{bmatrix} 2.8893 \\ 7.1107 \end{bmatrix} = \begin{bmatrix} 3 \\ 7 \end{bmatrix}.$$

3.3.18. Exercises 3.3

Let $A=\begin{bmatrix}4 & 2 & 0\\ 1 & -3 & 5\end{bmatrix}$, $B=\begin{bmatrix}2 & 4\\ -1 & 2\\ 3 & 5\end{bmatrix}$, $C=\begin{bmatrix}5 & 1\\ -2 & 3\end{bmatrix}$, $D=\begin{bmatrix}0 & 1\\ 2 & -3\end{bmatrix}$, $E=\begin{bmatrix}3\\ 1\\ 2\end{bmatrix}$, and $F=\begin{bmatrix}-2 & 4 & 1\end{bmatrix}$. In Exercises 1–12, compute the values of the matrix expressions, if they exist. If an expression is not defined, state the reason.

1. $AB+2C$.
2. $2AB-D^2$.
3. BAE.
4. $A+BC$.
5. FE.
6. EF.
7. $3C-2D$.
8. B^2.
9. C^2-D^3.
10. $A'+2B$.
11. $A-2B'$.
12. $A'C+B$.
13. Let $A=\begin{bmatrix}2 & -6 & 4\\ 7 & 1 & 0\\ -3 & 5 & 2\end{bmatrix}$, $B=\begin{bmatrix}4 & 1 & -2\\ 5 & 2 & 6\\ 0 & 4 & -3\end{bmatrix}$, and $C=\begin{bmatrix}9 & 2 & 1\\ -4 & 0 & 5\\ 3 & 4 & 2\end{bmatrix}$. Let O_3 and I_3 be zero and identity matrices of size 3. Verify the following:

   A. $A+O_3=O_3+A=A$
   B. $B-B=O_3$
   C. $B+C=C+B$
   D. $A+(B+C)=(A+B)+C$
   E. $CO_3=O_3C=O_3$
   F. $AI_3=I_3A=A$

14. Use MATLAB to verify that for any $n\times n$ matrices A, B, and C the following are true. Let n be a specific value, say $n=3$. Create the square matrices A, B, and C of size 3 with random integer entries between -10 and 10.

   A. $AI_n=I_nA=A$
   B. $AO_n=O_nA=O_n$

C. $A+O=O+A=A$

D. $A-A=O$

E. $A+B=B+A$

F. $A+(B+C)=(A+B)+C$

G. $A(BC)=(AB)C$

H. $A(B+C)=AB+AC$

Hint: Recall the following MTLAB predefined functions:

rand(n, n)	creates an $n \times n$ matrix in which the elements are real numbers between 0 and 1
2*rand(n, n) – 1	creates an $n \times n$ matrix in which the elements are real numbers between −1 and 1
round(10*(2*rand(n, n) – 1)	creates an $n \times n$ matrix in which the elements are integers between −10 and 10.

In Exercises 15–18,

A. Compute the inverse matrix if it exists using Gauss-Jordan elimination method.

B. Use MATLAB to find the inverse matrix

15. $A=\begin{bmatrix} -4 & 9 \\ -3 & 7 \end{bmatrix}.$

16. $A=\begin{bmatrix} 1 & 2 \\ 3 & 5 \end{bmatrix}.$

17. $A=\begin{bmatrix} -1 & 3 & 3 \\ 1 & -2 & -2 \\ -2 & -5 & -4 \end{bmatrix}.$

18. $A=\begin{bmatrix} 3 & 0 & 0 \\ 4 & 2 & 4 \\ -6 & 2 & 5 \end{bmatrix}.$

In Exercises 19–23,

A. Write the system in matrix form $AX=B$.

B. Solve the system by finding A^{-1} and evaluate $A^{-1}B$.

Hint: See Exercises 15–18.

19. $-4x_1+9x_2=2.$

$-3x_1+7x_2=2.$

20. $x_1 + 2x_2 = 5.$

$3x_1 + 5x_2 = 11.$

21. $-x_1 + 3x_2 + 3x_3 = -13.$

$x_1 - 2x_2 - 2x_3 = 10.$

$-2x_1 - 5x_2 - 4x_3 = 9.$

22. $3x_1 = -18.$

$4x_1 + 2x_2 + 4x_3 = -16.$

$-6x_1 + 2x_2 + 5x_3 = 45.$

23. $2x_1 + 3x_2 = 1.$

$-8x_1 - 10x_2 - x_3 = -12.$

$9x_1 + 11x_2 + x_3 = 14.$

In Exercises 24 and 25, consider Models 3.6 and 3.7, in which the population movement between a city and its surrounding suburbs was modeled by the matrix difference equation $X_n = TX_{n-1}$, where $T = \begin{bmatrix} 0.94 & 0.02 \\ 0.06 & 0.98 \end{bmatrix}$ is the transition matrix and $X_n = \begin{bmatrix} C_n \\ S_n \end{bmatrix}$ is the population vector after n years. Use MATLAB.

24. If $X_0 = \begin{bmatrix} 24{,}000 \\ 76{,}000 \end{bmatrix}$, find X_5.

25. If $X_6 = \begin{bmatrix} 32{,}568 \\ 67{,}432 \end{bmatrix}$, find X_0.

3.4. DETERMINANTS AND SYSTEMS OF LINEAR EQUATIONS

Every square matrix is associated with a real number called the **determinant of the matrix**. In this section, we introduce the main concepts of determinants and discuss a method to calculate them. The value of a determinant can be used to conclude whether a system of linear equations has a unique solution, many solutions, or no solution. The value of a determinant determines whether the matrix is invertible or not. If the matrix is invertible, the determinant is used in a formula to find the inverse of the matrix. We are interested in determinants because they play an important role in calculating the eigenvalues, which we investigate in Section 3.5.

Every square matrix is associated with a unique real number called its **determinant**.

Let's start with a 2×2 matrix:

$$A = \begin{bmatrix} a_{11} & a_{12} \\ a_{21} & a_{22} \end{bmatrix}.$$

The determinant of A, denoted by $|A|$ or det(A), is given by

$$|A| = \det(A) = a_{11}a_{22} - a_{12}a_{21}.$$

Note that the determinant of a 2×2 matrix is given by the difference of the product of the two diagonals.

Example 3.16

Find the determinant of the following matrices.

i. $A = \begin{bmatrix} 3 & 5 \\ -4 & 2 \end{bmatrix}.$

ii. $B = \begin{bmatrix} 0 & 4 \\ 2 & -3 \end{bmatrix}.$

iii. $C = \begin{bmatrix} 1/2 & 8 \\ 1/4 & 10 \end{bmatrix}.$

iv. $D = \begin{bmatrix} 1/2 & 0.25 \\ 7 & -5 \end{bmatrix}.$

Solution

i. $\det(A) = 3(2) - 5(-4) = 6 + 20 = 26.$
ii. $|B| = 0(-3) - 4(2) = 0 - 8 = -8.$
iii. $\det(C) = (1/2)(10) - 8(1/4) = 5 - 2 = 3.$
iv. $|D| = (1/2)(-5) - 0.25(7) = -2.50 - 1.75 = -4.25.$

The determinant of a 3×3 matrix is defined in terms of 2×2 matrices. The determinant of an $n \times n$ matrix is defined in terms of $(n-1) \times (n-1)$ matrices.

To define the determinant of a square matrix higher than 2, we need to define the terms **minor** and **cofactor**.

Definitions

If A is a square matrix, then the **minor** of the element a_{ij}, denoted by M_{ij}, is the determinant of the matrix obtained after deleting row i and column j of A. The **cofactor** of a_{ij}, denoted by C_{ij}, is given by

$$C_{ij} = (-1)^{i+j} M_{ij}.$$

Example 3.17

Find the minors and cofactors of the elements a_{11}, a_{23}, and a_{32} of matrix A:

$$A = \begin{bmatrix} 2 & 1 & 0 \\ -3 & 0 & 4 \\ 2 & -4 & 1 \end{bmatrix}.$$

Solution

The minor of a_{11}: $M_{11} = \begin{vmatrix} 0 & 4 \\ -4 & 1 \end{vmatrix} = 0 \times 1 - 4 \times (-4) = 16.$

The cofactor of a_{11}: $C_{11} = (-1)^{1+1} M_{11} = (-1)^2(16) = 16.$

The minor of a_{23}: $M_{23} = \begin{vmatrix} 2 & 1 \\ 2 & -4 \end{vmatrix} = 2 \times (-4) - 1 \times 2 = -10.$

The cofactor of a_{23}: $C_{23} = (-1)^{2+3} M_{23} = (-1)^5(-10) = 10.$

The minor of a_{32}: $M_{32} = \begin{vmatrix} 2 & 0 \\ -3 & 4 \end{vmatrix} = 2 \times 4 - 0 \times (-3) = 8.$

The cofactor of a_{32}: $C_{32} = (-1)^{2+3} M_{32} = (-1)^5(8) = -8.$

Definition

Let A be an $n \times n$ square matrix of order 2 or more. The determinant of A is the sum of the products of the elements of the first row (or any row) and their cofactors.

$$\det(A) = a_{11}C_{11} + a_{12}C_{12} + a_{13}C_{13} + \ldots + a_{1n}C_{1n}.$$

This formula is called a **cofactor expansion across the first row** of A.

Example 3.18

Find the determinant of matrix A, A:

$$A = \begin{bmatrix} 2 & 1 & 0 \\ -3 & 0 & 4 \\ 2 & -4 & 1 \end{bmatrix}.$$

Solution

Note that matrix A is the same matrix given in Example 3.17. The cofactors of the entries of the first row are

$$C_{11} = 16.$$

$$C_{12} = (-1)^{1+2} \begin{vmatrix} -3 & 4 \\ 2 & 1 \end{vmatrix} = (-1)\times(-3\times 1 - 4\times 2) = 11.$$

$$C_{13} = (-1)^{1+3} \begin{vmatrix} -3 & 0 \\ 2 & -4 \end{vmatrix} = (+1)(-3\times -4 - 0\times 2) = 12.$$

Therefore,

$$\begin{aligned} \det(A) &= a_{11}C_{11} + a_{12}C_{12} + a_{13}C_{13} \\ &= 2(16) + 1(11) + 0(12) = 32 + 11 + 0 = 43. \end{aligned}$$

Note: We defined the determinant of a square matrix as an expansion by the factors in the first row. It can be shown that the determinant can be found using any row or column. These facts are stated in the following theorem (we omit its proof).

Theorem

Let A be a $n \times n$ square matrix. The determinant of A is the sum of the products of the elements of any row or column and their cofactors. The ith row expansion is

$$\det(A) = a_{i1}C_{i1} + a_{i2}C_{i2} + \ldots + a_{in}C_{in},$$

and the jth column expansion is

$$\det(A) = a_{1j}C_{1j} + a_{2j}C_{2j} + \ldots + a_{nj}C_{nj}.$$

Example 3.19

Find the determinant of the matrix A using the second row and the third column:

$$A=\begin{bmatrix} 2 & 1 & 0 \\ -3 & 0 & 4 \\ 2 & -4 & 1 \end{bmatrix}.$$

Solution

To find det(A) using row 2, we have

$$\begin{aligned}
\det(A) &= a_{21}C_{21}+a_{22}C_{22}+a_{23}C_{23} \\
&= (-3)(-1)^{2+1}\begin{vmatrix} 1 & 0 \\ -4 & 1 \end{vmatrix}+(0)(-1)^{2+2}\begin{vmatrix} 2 & 0 \\ 2 & 1 \end{vmatrix}+(4)(-1)^{2+3}\begin{vmatrix} 2 & 1 \\ 2 & -4 \end{vmatrix} \\
&= (-3)(-1)(1\times 1-0\times(-4))+0+4(-1)(2\times(-4)-1\times 2) \\
&= 3+0+40=43.
\end{aligned}$$

To find det(A) using the column 3:

$$\det(A)=a_{13}C_{13}+a_{23}C_{23}+a_{33}C_{33}$$

$$\begin{aligned}
\det(A) &= (0)(-1)^{1+3}\begin{vmatrix} -3 & 0 \\ 2 & -4 \end{vmatrix}+4(-1)^{2+3}\begin{vmatrix} 2 & 1 \\ 2 & -4 \end{vmatrix}+1(-1)^{3+3}\begin{vmatrix} 2 & 1 \\ -3 & 0 \end{vmatrix} \\
&= 0+4(-1)(2\times(-4)-1\times 2)+1(1)(2\times 0-1\times(-3)) \\
&= 0+40+3=43.
\end{aligned}$$

3.4.1. Determinants in MATLAB

Let A be a square matrix. MATLAB has the function **det(A)** is used to calculate and return the determinant of A.

Consider the matrix A in Example 3.18:

```
>> A=[2 1 0; -3 0 4; 2 -4 1];
>> det(A)
ans =
  43
```

Definition

Let A be a square matrix. A is called **singular** if $\det(A)=0$ and **nonsingular** if $\det(A)\neq 0$.

Definition

Let A be an $n\times n$ matrix and C_{ij} be the cofactor of a_{ij}. The **matrix of cofactors** of A is the matrix

$$\begin{bmatrix} C_{11} & C_{12} & \cdots & C_{1n} \\ C_{21} & C_{22} & \cdots & C_{2n} \\ \vdots & \vdots & & \vdots \\ C_{n1} & C_{n2} & \cdots & C_{nn} \end{bmatrix},$$

where the (i, j)th element is the cofactor C_{ij} of a_{ij}.

The **adjoint** of A, denoted by $adj(A)$, is the transpose of the matrix of cofactors

$$adj(A) = \begin{bmatrix} C_{11} & C_{12} & \cdots & C_{1n} \\ C_{21} & C_{22} & \cdots & C_{2n} \\ \vdots & \vdots & & \vdots \\ C_{n1} & C_{n2} & \cdots & C_{nn} \end{bmatrix}^t.$$

Example 3.20

Find the matrix of cofactors and the adjoint matrix of the matrix A,

$$A = \begin{bmatrix} -1 & 3 & 3 \\ 1 & -2 & -2 \\ 2 & -5 & -4 \end{bmatrix}.$$

Solution

The cofactors of matrix A are

$$C_{11} = \begin{vmatrix} -2 & -2 \\ -5 & -4 \end{vmatrix} = -2, \quad C_{12} = -\begin{vmatrix} 1 & -2 \\ 2 & -4 \end{vmatrix} = 0, \quad C_{13} = \begin{vmatrix} 1 & -2 \\ 2 & -5 \end{vmatrix} = -1.$$

$$C_{21} = -\begin{vmatrix} 3 & 3 \\ -5 & -4 \end{vmatrix} = -3, \quad C_{22} = \begin{vmatrix} -1 & 3 \\ 2 & -4 \end{vmatrix} = -2, \quad C_{23} = -\begin{vmatrix} -1 & 3 \\ 2 & -5 \end{vmatrix} = 1.$$

$$C_{31} = \begin{vmatrix} 3 & 3 \\ -2 & -2 \end{vmatrix} = 0, \quad C_{32} = -\begin{vmatrix} -1 & 3 \\ 1 & -2 \end{vmatrix} = -1, \quad C_{33} = \begin{vmatrix} -1 & 3 \\ 1 & -2 \end{vmatrix} = -1.$$

The matrix of the cofactors of A is

$$\begin{bmatrix} -2 & 0 & -1 \\ -3 & -2 & 1 \\ 0 & 1 & -1 \end{bmatrix}.$$

We have

$$adj(A) = \begin{bmatrix} -2 & 0 & -1 \\ -3 & -2 & 1 \\ 0 & 1 & -1 \end{bmatrix}^t = \begin{bmatrix} -2 & -3 & 0 \\ 0 & -2 & 1 \\ -1 & 1 & -1 \end{bmatrix}.$$

Now, we state (without proof) another method of finding the inverse of a nonsingular square matrix.

Theorem

For a square matrix A if $\det(A) \neq 0$, then A is invertible and $A^{-1} = \frac{1}{\det(A)} adj(A)$.

Example 3.21

Let

$$A = \begin{bmatrix} -1 & 3 & 3 \\ 1 & -2 & -2 \\ 2 & -5 & -4 \end{bmatrix}.$$

$$B = \begin{bmatrix} 1 & 1 & -1 \\ -2 & -1 & 4 \\ 3 & 4 & -1 \end{bmatrix}.$$

i. Use the determinant to determine whether the matrix is invertible.
ii. If the matrix is invertible, use the formula for the inverse of a matrix to calculate the inverse of the matrix.

Solution

i. We have

$$\det(A) = (-1)(-1)^{1+1} \begin{vmatrix} -2 & -2 \\ -5 & -4 \end{vmatrix} + (3)(-1)^{1+2} \begin{vmatrix} 1 & -2 \\ 2 & -4 \end{vmatrix} + 3(-1)^{1+3} \begin{vmatrix} 1 & -2 \\ 2 & -5 \end{vmatrix}$$
$$= 2 + 0 - 3 = -1.$$

Because $\det(A) \neq 0$, matrix A is invertible.

For matrix B, we have

$$\det(B) = 1(-1)^{1+1}\begin{vmatrix} -1 & 4 \\ 4 & -1 \end{vmatrix} + (1)(-1)^{1+2}\begin{vmatrix} -2 & 4 \\ 3 & -1 \end{vmatrix} (-1)(1)^{1+3}\begin{vmatrix} -2 & -1 \\ 3 & 4 \end{vmatrix}$$

$$= -15 + 10 + 5 = 0.$$

Because $\det(B)=0$, matrix B is singular. Consequently, the inverse of B does not exist.

ii. We have

$$A^{-1} = \frac{1}{\det(A)} adj(A).$$

From Example 3.20 we have

$$adj(A) = \begin{bmatrix} -2 & -3 & 0 \\ 0 & -2 & 1 \\ -1 & 1 & -1 \end{bmatrix}.$$

Therefore,

$$A^{-1} = -\begin{bmatrix} -2 & -3 & 0 \\ 0 & -2 & 1 \\ -1 & 1 & -1 \end{bmatrix} = \begin{bmatrix} 2 & 3 & 0 \\ 0 & 2 & -1 \\ 1 & -1 & 1 \end{bmatrix}.$$

Let's use MATLAB to verify these results.

```
>> A=[-1 3 3; 1 -2 -2; 2 -5 -4]; B=[1 1 -1; -2 -1 4; 3 4 -1];
>> det(A)
ans =
  -1

>> det(B)
ans =
  5.5511e-16                          % Note that 5.5511e–16 = 5.5511/10^16 ≅ 0

>> inv(A)
ans =
  2  3  0
  0  2 -1
  1 -1  1
>> inv(B)
Warning: Matrix is close to singular or badly scaled.
Results may be inaccurate.
RCOND = 3.083953e-18.
```

```
ans =
   1.0e+16 *
  -2.7022 -0.5404  0.5404
   1.8014  0.3603 -0.3603
  -0.9007 -0.1801  0.1801
```

Note that MATLAB warning means that det(B) $\cong$ 0. Therefore inv(B) does not exist. The entries of inv(B) are $\cong$ zeros.

3.4.2. Determinants and Systems of Linear Equations

Given a system of n linear equations in n variables, we discuss the use of the determinant of the matrix of coefficients of the system to determine the existence and uniqueness of solutions to the system.

Assume we have the following system:

$$\begin{aligned} &a_{11}x_1 + a_{12}x_2 + \ldots + a_{1n}x_n = b_1 \\ &a_{21}x_1 + a_{22}x_2 + \ldots + a_{2n}x_n = b_2. \\ &\vdots \\ &a_{n1}x_1 + a_{n2}x_2 + \ldots + a_{nn}x_n = b_n \end{aligned}$$

This system can be expressed in the form

$$AX = B,$$

where

$$A = \begin{bmatrix} a_{11} & a_{12} & \cdots & a_{1n} \\ a_{21} & a_{22} & \cdots & a_{2n} \\ \vdots & \vdots & & \vdots \\ a_{n1} & a_{n2} & \cdots & a_{nn} \end{bmatrix}.$$

$$X = \begin{bmatrix} x_1 \\ x_2 \\ \vdots \\ x_n \end{bmatrix}.$$

$$B = \begin{bmatrix} b_1 \\ b_2 \\ \vdots \\ b_n \end{bmatrix}.$$

If A is invertible, multiply both sides of the equation by A^{-1} and we have

$$A^{-1}AX = A^{-1}B.$$
$$IX = A^{-1}B.$$
$$X = A^{-1}B.$$

Consequently, the system of n equations in n variables has the unique solution $X=A^{-1}B$. If $\det(A)=0$, then A^{-1} does not exist. Therefore, the system has many solutions or no solution.

Now consider the following homogenous system of equations

$$AX = O.$$

If $\det(A) \neq 0$, A^{-1} exists. Multiply both sides of the equation by A^{-1}, to get

$$A^{-1}AX = A^{-1}O.$$
$$IX = O.$$
$$X = O.$$

$X=O$ is called the **trivial solution**. Usually we are interested in nontrivial solutions. It is clear that to have a nontrivial solution to the homogenous system $AX=O$, we must have $\det(A)=0$.

Example 3.22

Determine whether the following homogenous system has a nontrivial solution. Find the solution if the system has one.

$$\begin{aligned} x_1 + \quad x_2 + 2x_3 &= 0. \\ -4x_1 \qquad\quad + 8x_3 &= 0. \\ 3x_1 + 2x_2 + 2x_3 &= 0. \end{aligned}$$

Solution

The matrix of coefficients A is

$$A = \begin{bmatrix} 1 & 1 & 2 \\ -4 & 0 & 8 \\ 3 & 2 & 2 \end{bmatrix}.$$

We have $\det(A)=0$. This implies that the homogenous system has a nontrivial solution.

Note that in Example 3.8, we showed that this system has many solutions in the form $x_1=2r$, $x_2=-4r$, and $x_3=r$, where r is a parameter.

Example 3.23

Find values of λ such that the following homogenous system of equations has nontrivial solutions. For each value of λ determine the solutions of the system.

$$(\lambda+3)x_1+6x_2=0$$
$$x_1+(\lambda-2)x_2=0.$$

Solution

The system can be represented by the matrix equation $AX=O$, where

$$A=\begin{bmatrix} \lambda+3 & 6 \\ 1 & \lambda-2 \end{bmatrix}.$$
$$X=\begin{bmatrix} x_1 \\ x_2 \end{bmatrix}.$$
$$O=\begin{bmatrix} 0 \\ 0 \end{bmatrix}.$$

The homogenous system has a nontrivial solution if $\det(A)=0$:

$$\begin{vmatrix} \lambda+3 & 6 \\ 1 & \lambda-2 \end{vmatrix}=0.$$
$$(\lambda+3)(\lambda-2)-6=0.$$
$$\lambda^2+\lambda-6-6=0.$$
$$\lambda^2+\lambda-12=0.$$
$$(\lambda+4)(\lambda-3)=0.$$

Therefore, $\lambda=-4$ or $\lambda=3$.

For $\lambda=-4$, the homogenous system is

$$(-4+3)x_1+6x_2=0.$$
$$x_1+(-4-2)x_2=0.$$

This can be simplified to

$$-x_1+6x_2=0$$
$$x_1-6x_2=0.$$

The reduced echelon form of augmented matrix

$$\begin{bmatrix} -1 & 6 & 0 \\ 1 & -6 & 0 \end{bmatrix}$$

is

$$\begin{bmatrix} 1 & -6 & 0 \\ 0 & 0 & 0 \end{bmatrix}.$$

Consequently,

$$\begin{aligned} x_1 - 6x_2 &= 0 \\ x_1 &= 6x_2. \end{aligned}$$

Therefore, for $\lambda = -4$, the system has many solutions in the form $x_1 = 6r$, $x_2 = r$, where r is a parameter, or $X = r\begin{bmatrix} 6 \\ 1 \end{bmatrix}$. Similarly for $\lambda = 3$, the system is

$$\begin{aligned} 6x_1 + 6x_2 &= 0 \\ x_1 + x_2 &= 0. \end{aligned}$$

This system has many solutions in the form $x_1 = -t$, $x_2 = t$, where t is a parameter. Equivalently, the system has many solutions in the form

$$X = t\begin{bmatrix} -1 \\ 1 \end{bmatrix}.$$

3.4.3. Exercises 3.4

1. Compute the determinant of each matrix and find the inverse, if it exists.

A. $A = \begin{bmatrix} 2 & 4 \\ 3 & 5 \end{bmatrix}$

B. $B = \begin{bmatrix} 2 & -4 \\ -1 & 2 \end{bmatrix}$

C. $C = \begin{bmatrix} 4 & 6 \\ 3 & 2 \end{bmatrix}$

D. $D = \begin{bmatrix} 7 & 14 \\ 2 & 4 \end{bmatrix}$

2. Compute the determinant of each matrix and find the inverse, if it exists.

A. $A = \begin{bmatrix} 3 & 2 \\ 4 & -5 \end{bmatrix}$

B. $B = \begin{bmatrix} -4 & 9 \\ -3 & 7 \end{bmatrix}$

C. $C = \begin{bmatrix} 5 & 4 \\ -3 & -2 \end{bmatrix}$

D. $D = \begin{bmatrix} 3/4 & 6 \\ 5/2 & 4 \end{bmatrix}$

3. Evaluate the determinant and the inverse of A if it exists.

$$A = \begin{bmatrix} 2 & 0 & 0 \\ 4 & 2 & 4 \\ -6 & 2 & 5 \end{bmatrix}$$

4. Let

$$A = \begin{bmatrix} 4 & 2 & 5 \\ 1 & 3 & 6 \\ 0 & 2 & 4 \end{bmatrix}$$

A. Evaluate det(A).

B. Find the inverse A^{-1}, if it exists, using the formula $A^{-1} = \dfrac{adj(A)}{\det(A)}$.

5. Let $A = \begin{bmatrix} -1 & 3 & 3 \\ 1 & -2 & -2 \\ -2 & -5 & -4 \end{bmatrix}$.

A. Find the determinant of A.

B. Find the matrix of cofactors and the adjoint matrix of the matrix A.

C. Find the inverse matrix A^{-1}, if it exists, using the formula $A^{-1} = \dfrac{adj(A)}{\det(A)}$.

D. Use the result from part C to solve the following system:

$$\begin{aligned} -x_1 + 3x_2 + 3x_3 &= -8 \\ x_1 - 2x_2 - 2x_3 &= 6 \\ -2x_1 - 5x_2 - 4x_3 &= 3 \end{aligned}$$

E. Check your answers with MATLAB.

6. Let $A = \begin{bmatrix} 4 & 0 & -3 \\ 1 & 5 & 0 \\ -2 & 3 & 2 \end{bmatrix}$.

A. Find the determinant of A.
B. Find the matrix of cofactors and the adjoint matrix of A.
C. Find the inverse matrix A^{-1}, if it exists, using the formula $A^{-1} = \dfrac{adj(A)}{\det(A)}$.
D. Use the result from part C to solve the following system:

$$\begin{aligned} 4x_1 \qquad - 3x_3 &= -5 \\ x_1 + 5x_2 \qquad &= 13 \\ -2x_1 + 3x_2 + 2x_2 &= 11 \end{aligned}$$

E. Check your answers with MATLAB.

3.5. EIGENVALUES AND EIGENVECTORS

In this section, we introduce special scalars and vectors associated with square matrices called eigenvalues and eigenvectors. The eigenvalues and eigenvectors play an important role in mathematical modeling in many areas, including the life sciences. In this section we intuitively introduce the concept of eigenvalues and eigenvectors and investigate methods to determine the eigenvalues of a square matrix and the eigenvectors corresponding to each eigenvalue.

3.5.1. Exploration 3.1

Consider a species that can live for 2 years. We will group the population in age classes [0, 1) and [1, 2]. Assume that the species has the following characteristics:

1. The survival rate of the age classes [0, 1) and [1, 2] are 0.4 and 0, respectively.
2. The average number of offspring produced by the [0, 1) and [1, 2] age classes are 1 and 5, respectively.

Let a_n and b_n be the population of the species in the age classes [0, 1) and [1, 2] after n years, respectively. Assume that the initial populations are $a_0 = 5$ and $b_0 = 5$.

i. Find a_n and b_n for $n = 1$, 2, and 3.
ii. Model this situation by a system of difference equations.
iii. Letting $X_n = \begin{bmatrix} a_n \\ b_n \end{bmatrix}$, model this situation by a single matrix equation. Find the analytical solution of the matrix equation.
iv. Use the matrix equation developed in part iii to find a_n and b_n for $n = 0, 1, 2, \ldots, 10$. Do you see a pattern in the vectors $\begin{bmatrix} a_n \\ b_n \end{bmatrix}$? If there is a pattern describe it.
v. Repeat part iv with the initial value $X_0 = \begin{bmatrix} 5 \\ 1 \end{bmatrix}$.

Discussion

i. Using the given survival and fertility rates and the initial population, we can calculate the population at the end of the first three periods. We have

$$a_1 = 30 \text{ and } b_1 = 2.$$
$$a_2 = 40 \text{ and } b_2 = 12.$$
$$a_3 = 100 \text{ and } b_3 = 16.$$

ii. This situation can be modeled by the following system of first-order linear difference equations:

$$a_{n+1} = 1 \cdot a_n + 5 \cdot b_n.$$
$$b_{n+1} = 0.4a_n.$$

iii. This system of equations can be represented by the following matrix first-order difference equation

$$X_{n+1} = TX_n, \quad n = 0, 1, 2, \ldots,$$

where

$$X_n = \begin{bmatrix} a_n \\ b_n \end{bmatrix}, \quad T = \begin{bmatrix} 1 & 5 \\ 0.4 & 0 \end{bmatrix}, \quad X_0 = \begin{bmatrix} 5 \\ 5 \end{bmatrix}.$$

The analytical solution of this equation is

$$X_n = T^n X_0.$$

iv. You may use the following MATLAB code to get X_n, $n=0, 1, 2, \dots, 10$.

```
X=[5; 5];
T=[1 5; 0.4 0];
M=[X];
for k=1:10
  X=T*X;
  M=[M X];
end;
M
```

The values of X_n, $n=0,1,2,\dots,10$ are $\begin{bmatrix}5\\5\end{bmatrix}, \begin{bmatrix}30\\2\end{bmatrix}, \begin{bmatrix}40\\12\end{bmatrix}, \begin{bmatrix}100\\16\end{bmatrix}, \begin{bmatrix}180\\40\end{bmatrix}, \begin{bmatrix}380\\72\end{bmatrix},$ $\begin{bmatrix}740\\152\end{bmatrix}, \begin{bmatrix}1500\\296\end{bmatrix}, \begin{bmatrix}2980\\600\end{bmatrix}, \begin{bmatrix}2560\\512\end{bmatrix}$, and $\begin{bmatrix}11940\\2392\end{bmatrix}$. There is no pattern in the vectors $\begin{bmatrix}a_n\\b_n\end{bmatrix}$.

v. Similar to part iv the values of X_n, $n=0, 1, 2, \dots, 10$ are $\begin{bmatrix}5\\1\end{bmatrix}, \begin{bmatrix}10\\2\end{bmatrix}, \begin{bmatrix}20\\4\end{bmatrix}, \begin{bmatrix}40\\8\end{bmatrix},$ $\begin{bmatrix}80\\16\end{bmatrix}, \begin{bmatrix}160\\32\end{bmatrix}, \begin{bmatrix}320\\64\end{bmatrix}, \begin{bmatrix}640\\128\end{bmatrix}, \begin{bmatrix}1280\\256\end{bmatrix}, \begin{bmatrix}2560\\512\end{bmatrix}$, and $\begin{bmatrix}5120\\1024\end{bmatrix}$. There is a pattern in the dynamic of the vectors $\begin{bmatrix}a_n\\b_n\end{bmatrix}$. The population of each class increases with the same ratio—that is, $a_{n+1}=2a_n$ and $b_{n+1}=2b_n$. This fact can be expressed by the following equation:

$$X_{n+1} = 2X_n.$$

Another interpretation of the pattern is that the percentage of the population of each age class is the same. For instance, $a_n = \frac{5}{6}(a_n + b_n) = 0.8333(a_n + b_n)$ and $b_n = \frac{1}{6}(a_n + b_n) = 0.1667(a_n + b_n)$. Letting $p_n=(a_n+b_n)$, we have $a_n \cong$ 83.3% of p_n and $b_n \cong 16.7\%$ of p_n.

The equations $X_{n+1}=TX_n$ and $X_{n+1}=2X_n$ imply that $TX_n=2X_n$, or in general, $TX=2X$. In this equation, the scalar 2 is called an **eigenvalue** of the matrix T, and X is called an **eigenvector** of T, corresponding to the eigenvalue 2.

3.5.2. Eigenvalues and Eigenvectors

Let A be an $n \times n$ matrix. A scalar λ is called an **eigenvalue** of A if there exists a nonzero $n \times 1$ column vector X such that

$$AX = \lambda X \tag{3.4}$$

X is called an **eigenvector** of A corresponding to λ. In practice, eigenvalues of a matrix are found first and then the corresponding eigenvectors are found. Equation 3.4 can be rewritten as

$$(A - \lambda I)X = O,$$

which represents a system of homogenous linear equations, where I is $n \times n$ identity matrix and O is the $n \times 1$ zero matrix. This system of equations has nonzero solutions if and only if the determinant of the coefficient matrix $(A - \lambda I)$ is zero—that is,

$$\det(A - \lambda I) = 0 \tag{3.5}$$

The expression $\det(A - \lambda I)$ is a polynomial in λ and is called the **characteristic polynomial** of A. The roots of the characteristic polynomial of A are the eigenvalues of A. For each eigenvalue, λ_i, we find the corresponding eigenvectors by solving the homogeneous system of equations $(A - \lambda_i I)X = O$ for X.

We will give a geometric interpretation of an eigenvalue, λ, of matrix A and a corresponding eigenvector **u**. Because $A\mathbf{u} = \lambda\mathbf{u}$, multiplication of **u** by matrix A produces vector $\lambda\mathbf{u}$. If $\lambda > 0$, vector $\lambda\mathbf{u}$ is in the same direction as **u**, and if $\lambda < 0$, vector $\lambda\mathbf{u}$ is in the opposite direction of **u**.

For an example of a positive eigenvalue, assume that $\lambda = 3$ is an eigenvalue of matrix A and assume that $\mathrm{u} = \begin{bmatrix} 2 \\ 1 \end{bmatrix}$ is the eigenvector corresponding to $\lambda = 3$. We have

$$A\mathbf{u} = \lambda\mathbf{u}.$$

$$A\begin{bmatrix} 2 \\ 1 \end{bmatrix} = 3\begin{bmatrix} 2 \\ 1 \end{bmatrix} = \begin{bmatrix} 6 \\ 3 \end{bmatrix}.$$

As shown in Figure 3.1A, the vector $\begin{bmatrix} 6 \\ 3 \end{bmatrix}$ is in the same direction as the eigenvector $\begin{bmatrix} 2 \\ 1 \end{bmatrix}$, and the length of $\begin{bmatrix} 6 \\ 3 \end{bmatrix}$ is 3 times the length of the eigenvector $\begin{bmatrix} 2 \\ 1 \end{bmatrix}$.

For an example of a negative eigenvalue, assume that $\lambda = -2$ is an eigenvalue of matrix A, and assume that $\mathrm{u} = \begin{bmatrix} 2 \\ 1 \end{bmatrix}$ is the eigenvector corresponding to $\lambda = -2$. We have

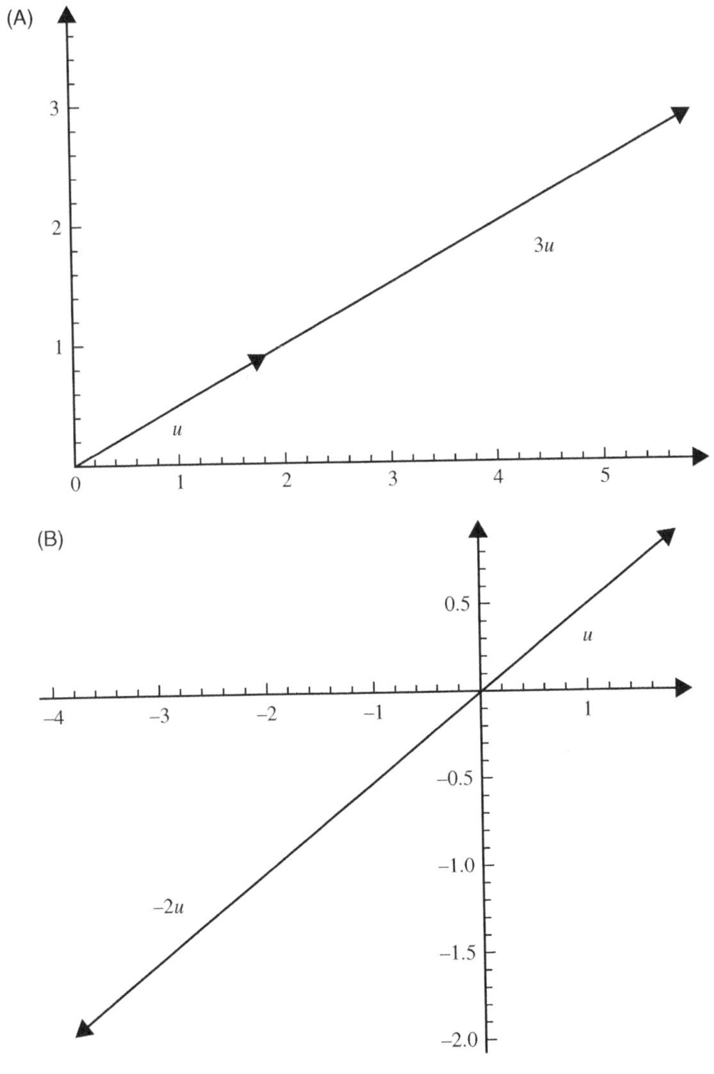

FIGURE 3.1. **A**. The vector $u = \begin{bmatrix} 2 \\ 1 \end{bmatrix}$ is an eigenvector of a matrix A corresponding to the eigenvalue $\lambda = 3$. $Au = \lambda u = 3u$ is a vector in the same direction as u and is three times the length of u. **B**. The vector $u = \begin{bmatrix} 2 \\ 1 \end{bmatrix}$ is an eigenvector of a matrix A corresponding to the eigenvalue $\lambda = -2$. $Au = \lambda u = -2u = \begin{bmatrix} -4 \\ -2 \end{bmatrix}$ is a vector in the opposite direction of u and is two times the length of u.

$$A\begin{bmatrix}2\\1\end{bmatrix}=-2\begin{bmatrix}2\\1\end{bmatrix}=\begin{bmatrix}-4\\-2\end{bmatrix}.$$

Figure 3.1B shows that the vector $\begin{bmatrix}-4\\-2\end{bmatrix}$ is in the opposite direction of the eigenvector $\begin{bmatrix}2\\1\end{bmatrix}$ and the length of $\begin{bmatrix}-4\\-2\end{bmatrix}$ is 2 times the length of the eigenvector $\begin{bmatrix}2\\1\end{bmatrix}$.

Example 3.24

Let $A=\begin{bmatrix}2 & 1\\-5 & -4\end{bmatrix}$.

i. Show that $X=\begin{bmatrix}-1\\5\end{bmatrix}$ is an eigenvector of A. Find the corresponding eigenvalue.

ii. Show that $Y=\begin{bmatrix}2\\3\end{bmatrix}$ is not an eigenvector of A.

Solution

i. The vector X is an eigenvector of A corresponding to the eigenvalue λ if $AX=\lambda X$. We have

$$AX=\begin{bmatrix}2 & 1\\-5 & -4\end{bmatrix}\begin{bmatrix}-1\\5\end{bmatrix}=\begin{bmatrix}3\\-15\end{bmatrix}=-3\begin{bmatrix}-1\\5\end{bmatrix}=-3X.$$

Therefore $X=\begin{bmatrix}-1\\5\end{bmatrix}$ is an eigenvector of A corresponding to the eigenvalue -3.

ii. $AY=\begin{bmatrix}2 & 1\\-5 & -4\end{bmatrix}\begin{bmatrix}2\\3\end{bmatrix}=\begin{bmatrix}7\\-22\end{bmatrix}\neq\lambda\begin{bmatrix}2\\3\end{bmatrix}$ for any λ.

Because AY is not a multiple of Y, vector Y is not an eigenvector of A.

Example 3.25

Let $A=\begin{bmatrix}2 & 5\\6 & 1\end{bmatrix}$. Show that 7 is an eigenvalue of A. Determine the eigenvectors corresponding to the eigenvalue 7.

Solution

We need to show that $AX=7X$ for some nonzero vector X. We have

$$AX = 7X.$$
$$AX - 7IX = O.$$
$$(A - 7I)X = O.$$
$$\left(\begin{bmatrix} 2 & 5 \\ 6 & 1 \end{bmatrix} - 7\begin{bmatrix} 1 & 0 \\ 0 & 1 \end{bmatrix}\right)\begin{bmatrix} x_1 \\ x_2 \end{bmatrix} = \begin{bmatrix} 0 \\ 0 \end{bmatrix}.$$
$$\begin{bmatrix} -5 & 5 \\ 6 & -6 \end{bmatrix}\begin{bmatrix} x_1 \\ x_2 \end{bmatrix} = \begin{bmatrix} 0 \\ 0 \end{bmatrix}.$$

The augmented matrix is

$$\begin{bmatrix} -5 & 5 & 0 \\ 6 & -6 & 0 \end{bmatrix}.$$

The reduced echelon form is

$$\begin{bmatrix} 1 & -1 & 0 \\ 0 & 0 & 0 \end{bmatrix}.$$

Therefore,

$$x_1 - x_2 = 0.$$
$$x_1 = x_2.$$

Letting $x_2 = r$ (a parameter), vector X is

$$X = r\begin{bmatrix} 1 \\ 1 \end{bmatrix}.$$

Because a nonzero vector X exists such that $AX = 7X$, 7 is an eigenvalue of A, and the corresponding eigenvectors are $r\begin{bmatrix} 1 \\ 1 \end{bmatrix}$.

In Examples 3.24 and 3.25, we were able to check whether a given vector was an eigenvector and if a given scalar was an eigenvalue. The interesting task is to find the eigenvalues and the corresponding eigenvectors for a given matrix. A few examples follow.

Example 3.36

Find the eigenvalues and the corresponding eigenvectors of matrix A considered in Exploration 3.1:

$$A = \begin{bmatrix} 1 & 5 \\ 0.4 & 0 \end{bmatrix}.$$

Solution

Step 1: Find the characteristic polynomial of matrix A, $\det(A-\lambda I_2)$.

We have

$$A-\lambda I_2 = \begin{bmatrix} 1 & 5 \\ 0.4 & 0 \end{bmatrix} - \lambda \begin{bmatrix} 1 & 0 \\ 0 & 1 \end{bmatrix} = \begin{bmatrix} 1 & 5 \\ 0.4 & 0 \end{bmatrix} + \begin{bmatrix} -\lambda & 0 \\ 0 & -\lambda \end{bmatrix} = \begin{bmatrix} 1-\lambda & 5 \\ 0.4 & -\lambda \end{bmatrix}.$$

The characteristic polynomial of A is

$$\begin{aligned} \det(A-\lambda I_2) &= \det\begin{bmatrix} 1-\lambda & 5 \\ 0.4 & -\lambda \end{bmatrix} \\ &= (1-\lambda)(-\lambda) - 0.4(5) = \lambda^2 - \lambda - 2. \end{aligned}$$

Recall that $\det\begin{bmatrix} a & b \\ c & d \end{bmatrix} = ad - cb.$

Step 2: The eigenvalues of A are the solutions of the equation:

$$\det(A-\lambda I) = 0.$$

We have

$$\begin{aligned} \lambda^2 - \lambda - 2 &= 0. \\ (\lambda-2)(\lambda+1) &= 0. \\ (\lambda-2) = 0 \quad \text{or} \quad (\lambda+1) &= 0. \end{aligned}$$

Therefore, the eigenvalues of A are $\lambda_1 = 2$ and $\lambda_2 = -1$.

To find the eigenvectors corresponding to the eigenvalue λ_k, solve the following matrix equation:

$$(A-\lambda_k I)X = O$$

for X. To find the eigenvectors corresponding to the eigenvalue $\lambda_1 = 2$, we solve

$$(A-2I)X = O.$$

We have

$$\left(\begin{bmatrix} 1 & 5 \\ 0.4 & 0 \end{bmatrix} - 2\begin{bmatrix} 1 & 0 \\ 0 & 1 \end{bmatrix}\right)\begin{bmatrix} x_1 \\ x_2 \end{bmatrix} = \begin{bmatrix} 0 \\ 0 \end{bmatrix}.$$

$$\begin{bmatrix} -1 & 5 \\ 0.4 & -2 \end{bmatrix}\begin{bmatrix} x_1 \\ x_2 \end{bmatrix} = \begin{bmatrix} 0 \\ 0 \end{bmatrix}.$$

$$\begin{bmatrix} -x_1 + 5x_2 \\ 0.4x_1 - 2x_2 \end{bmatrix} = \begin{bmatrix} 0 \\ 0 \end{bmatrix}.$$

Thus

$$-x_1 + 5x_2 = 0.$$
$$0.4x_1 - 2x_2 = 0.$$

To solve this homogenous system of two linear equations, we use the following augmented matrix:

$$\begin{bmatrix} -1 & 5 & 0 \\ 0.4 & -2 & 0 \end{bmatrix}.$$

The reduced echelon form of the augmented matrix is

$$\begin{bmatrix} 1 & -5 & 0 \\ 0 & 0 & 0 \end{bmatrix}.$$

The corresponding system of equations is

$$x_1 - 5x_2 = 0.$$

That is,

$$x_1 = 5x_2.$$

Let $x_2 = r$ (a parameter). The solutions of the system of equations are

$$x_1 = 5r.$$
$$x_2 = r.$$

Therefore, the eigenvectors of A corresponding to $\lambda = 2$ are nonzero vectors of the form

$$X = r\begin{bmatrix} 5 \\ 1 \end{bmatrix}.$$

Note that for each value of the parameter r there is an eigenvector. For example, if $r = 3$, then $\begin{bmatrix} 15 \\ 3 \end{bmatrix}$ is an eigenvector.

To check that $V=\begin{bmatrix}5\\1\end{bmatrix}$ is an eigenvector corresponding to the eigenvalue $\lambda=2$, show that $AV=2V$. We have

$$AV=\begin{bmatrix}1 & 5\\0.4 & 0\end{bmatrix}\begin{bmatrix}5\\1\end{bmatrix}=\begin{bmatrix}10\\2\end{bmatrix}=2\begin{bmatrix}5\\1\end{bmatrix}=2V.$$

To find the eigenvectors corresponding to the eigenvalue $\lambda_2=-1$, we solve

$$(A-(-1)I)X=0.$$
$$(A+I)X=0.$$

We have

$$\left(\begin{bmatrix}1 & 5\\0.4 & 0\end{bmatrix}+\begin{bmatrix}1 & 0\\0 & 1\end{bmatrix}\right)\begin{bmatrix}x_1\\x_2\end{bmatrix}=\begin{bmatrix}0\\0\end{bmatrix}.$$

$$\begin{bmatrix}2 & 5\\0.4 & 1\end{bmatrix}\begin{bmatrix}x_1\\x_2\end{bmatrix}=\begin{bmatrix}0\\0\end{bmatrix}.$$

$$\begin{bmatrix}2x_1+5x_2\\0.4x_1+x_2\end{bmatrix}=\begin{bmatrix}0\\0\end{bmatrix}.$$

Thus

$$2x_1+5x_2=0.$$
$$0.4x_1+5x_2=0.$$

The reduced echelon form of the augmented matrix $\begin{bmatrix}2 & 5 & 0\\0.4 & 1 & 0\end{bmatrix}$ is

$$\begin{bmatrix}1 & 5/2 & 0\\0 & 0 & 0\end{bmatrix}.$$

The corresponding system of equations is

$$x_1+\frac{5}{2}x_2=0.$$

That is,

$$x_1 = -\frac{5}{2}x_2.$$

Let $x_2 = t$ (a parameter). The solutions of the system of equations are

$$x_1 = -\frac{5}{2}t.$$

$$x_2 = t.$$

Therefore, the eigenvectors of A corresponding to $\lambda = -1$ are vectors of the form

$$X = t\begin{bmatrix} -5/2 \\ 1 \end{bmatrix}.$$

Note that for each value of the parameter t there is an eigenvector. For example, if $t=2$, then $\begin{bmatrix} -5 \\ 2 \end{bmatrix}$ is an eigenvector.

Here are geometric interpretations of the eigenvalues and corresponding eigenvectors for this example. The eigenvector $u = \begin{bmatrix} 5 \\ 1 \end{bmatrix}$ is shown in Figure 3.2A. Since we have

$$Au = 2u = 2\begin{bmatrix} 5 \\ 1 \end{bmatrix} = \begin{bmatrix} 10 \\ 2 \end{bmatrix},$$

the vector Au is in the same direction as u and twice the length of u (Figure 3.2A).

The second eigenvector, $u = \begin{bmatrix} -5/2 \\ 1 \end{bmatrix}$, is shown in Figure 3.2B. Because we have

$$Au = -u = -\begin{bmatrix} -5/2 \\ 1 \end{bmatrix} = \begin{bmatrix} 5/2 \\ -1 \end{bmatrix},$$

the vector Au is in the opposite direction of u and has the same length as u (see Figure 3.2B).

Example 3.27: Repeated Real Eigenvalues

Find the eigenvalues and the corresponding eigenvectors of the following matrix:

$$A = \begin{bmatrix} 1 & 1 \\ -1 & 3 \end{bmatrix}$$

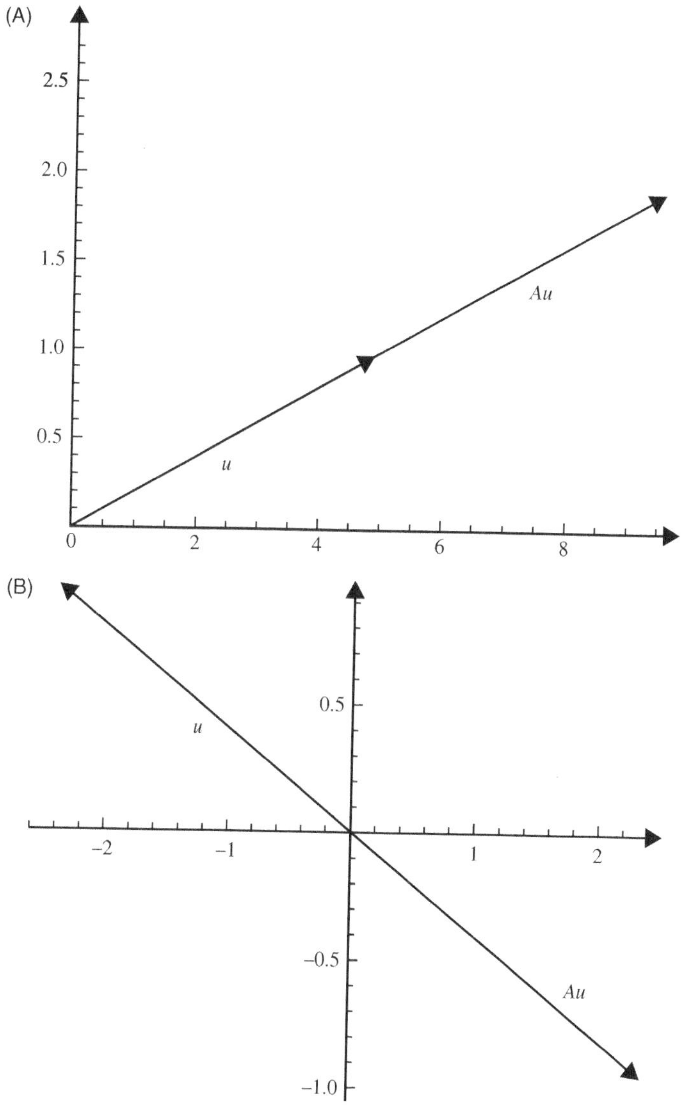

FIGURE 3.2. **A**. The vector $u = \begin{bmatrix} 5 \\ 1 \end{bmatrix}$ is an eigenvector of the matrix $A = \begin{bmatrix} 1 & 5 \\ 0.4 & 0 \end{bmatrix}$ corresponding to the eigenvalue $\lambda = 2$. The vector $Au = 2u$ is in the same direction as u and is two times the length of u. **B**. The vector $u = \begin{bmatrix} -5/2 \\ 1 \end{bmatrix}$ is an eigenvector of the matrix $A = \begin{bmatrix} 1 & 5 \\ 0.4 & 0 \end{bmatrix}$ corresponding to the eigenvalue $\lambda = -1$. The vector $Au = -u$ is in the opposite direction of u and has the same length as u.

Solution

First, find the characteristic polynomial of matrix A, $\det(A - \lambda I_2)$. We have

$$A - \lambda I_2 = \begin{bmatrix} 1 & 1 \\ -1 & 3 \end{bmatrix} - \lambda \begin{bmatrix} 1 & 0 \\ 0 & 1 \end{bmatrix} = \begin{bmatrix} 1-\lambda & 1 \\ -1 & 3-\lambda \end{bmatrix}.$$

The characteristic polynomial of A is

$$\det(A - \lambda I_2) = \det \begin{bmatrix} 1-\lambda & 1 \\ -1 & 3-\lambda \end{bmatrix}$$
$$= (1-\lambda)(3-\lambda) + 1 = \lambda^2 - 4\lambda + 4.$$

The eigenvalues of A are the solutions of the equation:

$$\lambda^2 - 4\lambda + 4 = 0.$$
$$(\lambda - 2)(\lambda - 2) = 0$$
$$(\lambda - 2) = 0 \quad \text{or} \quad (\lambda - 2) = 0.$$

Therefore, there is one repeated real eigenvalue, $\lambda = 2$, of matrix A. Because eigenvalue 2 is repeated twice, we say eigenvalue $\lambda = 2$ is of **multiplicity** 2.

To find the eigenvectors corresponding to the eigenvalue $\lambda = 2$, we solve

$$(A - 2I)X = O.$$

We have

$$\left(\begin{bmatrix} 1 & 1 \\ -1 & 3 \end{bmatrix} - 2 \begin{bmatrix} 1 & 0 \\ 0 & 1 \end{bmatrix} \right) \begin{bmatrix} x_1 \\ x_2 \end{bmatrix} = \begin{bmatrix} 0 \\ 0 \end{bmatrix}.$$

$$\begin{bmatrix} -1 & 1 \\ -1 & 1 \end{bmatrix} \begin{bmatrix} x_1 \\ x_2 \end{bmatrix} = \begin{bmatrix} 0 \\ 0 \end{bmatrix}.$$

To solve this homogenous system of two linear equations, we use the following augmented matrix:

$$\begin{bmatrix} -1 & 1 & 0 \\ -1 & 1 & 0 \end{bmatrix}.$$

The reduced echelon form of the augmented matrix is

$$\begin{bmatrix} 1 & -1 & 0 \\ 0 & 0 & 0 \end{bmatrix}.$$

Therefore, the eigenvectors of A corresponding to $\lambda = 2$ are nonzero vectors of the form

$$V = r \begin{bmatrix} 1 \\ 1 \end{bmatrix},$$

where r is a parameter.

3.5.3. Computing Eigenvalues and Eigenvectors with MATLAB

Let A be a square matrix of dimension n (i.e., $n \times n$ matrix). We discuss two methods for computing the eigenvalues and corresponding eigenvectors of the matrix A.

Method 1

The function **poly (A)** computes the characteristic polynomial of A. If A is an $n \times n$ matrix, **poly (A)** returns the coefficients of the characteristic polynomial in a row vector of length $n+1$.

The command **roots(C)** calculates the roots of the polynomial whose coefficients are stored in vector C.

The command **roots (poly(A))** returns the eigenvalues of matrix A. The output is a column vector.

The function **eig (A)** is equivalent to **roots (poly (A))** and returns the eigenvalues of matrix A in a column vector.

To find the eigenvector V corresponding to the eigenvalue λ of matrix A, you need to find the solution of the homogenous system $(A - \lambda I)V = O$.

The command **rref(A- λ*eye(n))** provides the reduced echelon form of the system, and consequently, from that form you can find the eigenvector(s).

Method 2.

As we discussed above **eig(A)** calculates the eigenvalues of matrix A. However, **eig (A)** can be used to calculate the eigenvalues and the corresponding eigenvectors of A if **eig (A)** is used in the form

```
[V D] = eig (A)
```

In this form, D is a diagonal matrix in which the eigenvalues of A are the entries along the diagonal. The columns of matrix V are the eigenvectors corresponding to the eigenvalues in D. That is, the ith column of V, $V(:, i)$, is the eigenvector corresponding to the eigenvalue $D(i, i)$.

Note that the eigenvectors in V are normalized. That is, the norm of each eigenvector (column) is 1. The general eigenvectors are scaled multiples of the normalized eigenvectors. Recall that the norm (length) of a vector $V = (v_1, v_2, \ldots, v_n)$ in $\mathbb{R}^n$ is defined by

$$\|V\| = \sqrt{(v_1)^2 + (v_2)^2 + \ldots + (v_n)^2}.$$

To normalize a vector V, multiply the vector by the scalar $\dfrac{1}{\|V\|}$.

Let's use these two methods to calculate the eigenvalues and the corresponding eigenvectors of matrix A, discussed in Example 3.26:

$$A = \begin{bmatrix} 1 & 5 \\ 0.4 & 0 \end{bmatrix}.$$

```
>> A=[1 5; 0.4 -4]
A=
  1.0000 5.0000
  0.4000      0
```

% Enter the matrix A

```
>> eig(A)
ans=
   2
  -1
```

% To find the eigenvalues of A

% $\lambda_1=2$
% $\lambda_2=-1$

```
>> rref (A -2*eye(2))
>>
ans=
  1 -5
  0  0
```

% To calculate the eigenvectors corresponding
% to $\lambda=2$

%This means that $x_1-5x_2=0 \Rightarrow x_1=5x_2$

```
>> rref (A -(-1)*eye(2))
>>
ans=
  1.0000 2.5000
       0      0
>> roots (poly (A))
>>
ans=
   2
  -1
```

% To calculate the eigenvectors corresponding
% to $\lambda=-1$

%This means that $x_1+2.5x_2=0 \Rightarrow x_1=-2.5x_2$

% Another way to calculate the eigenvalues
% of A

```
>>
```

% Calculation of eigenvalues and corresponding eigenvectors of A

```
>> [V D] =eig(A)
>>
V=
  0.9806 -0.9285
  0.1961  0.3714
D=
  2  0
  0 -1

>>  % The first column of V is the eigenvector associated with
>>  % the eigenvalue 2 (the first value of the diagonal of D)
>>  % Similarly the second column of V is the eigenvector
>>  % corresponding to the eigenvalue -1
```

Note that the eigenvector $V=\begin{bmatrix}0.9806\\0.1961\end{bmatrix}$ corresponding to the eigenvalue $\lambda=2$ is normalized—that is, $\|V\|=1$. The vector V can be written in the form

$$V=\begin{bmatrix}0.9806\\0.1961\end{bmatrix}=0.1961\begin{bmatrix}5\\1\end{bmatrix}.$$

This form is the same as $V=r\begin{bmatrix}5\\1\end{bmatrix}$, where r is a parameter.

Similarly the eigenvector associated with the eigenvalue $\lambda=-1$ is $V=\begin{bmatrix}-0.9285\\0.3714\end{bmatrix}$, which can be written in the form $V=\begin{bmatrix}-0.9806\\0.3714\end{bmatrix}=0.3714\begin{bmatrix}-2.5\\1\end{bmatrix}$. The general form of the eigenvectors corresponding to $\lambda=-1$ is $V=t\begin{bmatrix}-2.5\\1\end{bmatrix}$, where t is a parameter.

3.5.4. Complex Numbers

So far we have studied real distinct eigenvalues and real repeated eigenvalues. Sometimes there are no real solutions of the characteristic polynomial of a matrix. Before we study complex eigenvalues and complex eigenvectors let's review complex numbers.

Suppose, for example, we want to solve the quadratic equation $x^2+4=0$. We have

$$x^2=-4.$$

$$x=\pm\sqrt{-4}.$$

It is clear that there is no real number, whose square equals -4. So we say that there are no real solutions for the equation $x^2=-4$. However, we can write

$$\sqrt{-4}=\sqrt{4(-1)}=\sqrt{4}\sqrt{-1}=2\sqrt{-1}.$$

Mathematicians define $\sqrt{-1}$ as an **imaginary unit** i. Note that $i^2=\left(\sqrt{-1}\right)^2=-1$.

Let's review complex numbers. A complex number is a number of the form $a+bi$, where a and b are real numbers, and $i=\sqrt{-1}$ is the imaginary unit. The variable a is called the **real part**, and bi is called the **imaginary part** of the complex number. We will represent the complex number $a+bi$ by point (a, b) on a **complex plane**. The complex plane has two axes: the horizontal axis is the **real axis**, denoted by Re z, and the vertical axis is the **imaginary axis**, denoted by Im z. For example, Figure 3.3 shows the complex number $2+3i$ is represented on a complex plane.

Another way to represent a complex number $a+bi$ is as a vector in a complex plane, where the real (horizontal) coordinate is a and the imaginary (vertical) coordinate is b. For example, Figure 3.4 shows the complex numbers $2+3i$ and $-3-2i$ are on a complex plane.

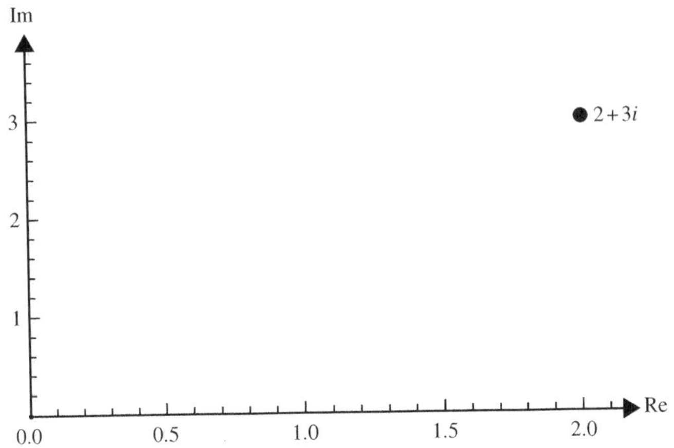

FIGURE 3.3. The complex number is $2 + 3i$ represented as a point in the complex plane.

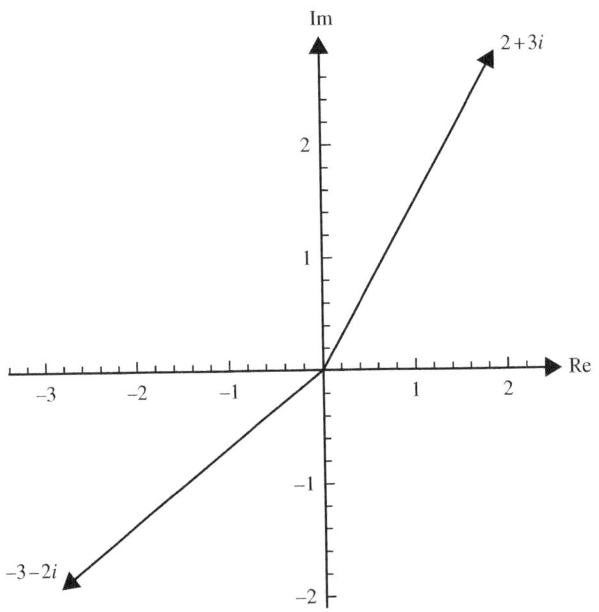

FIGURE 3.4. The complex numbers $2+3i$ and $-3-2i$ are represented as vectors in the complex plane.

Arithmetic of Complex Numbers

Scalar Multiplication

Let c be a real number and $z=a+bi$ be a complex number. We define

$$cz = c(a+bi) = ca + cbi,$$

and call it the **scalar multiple** of c and the complex number $z=a+bi$. For example, $2(3-4i)=6-8i$ and $-3(2+i)=-6-3i$.

Addition and Difference of Complex Numbers
Let $z=a+bi$ and $w=c+di$ be complex numbers. The **sum (addition)** of z and w is defined by

$$z+w=(a+bi)+(c+di)=(a+c)+(b+d)i.$$

The **difference** of z and w is defined by

$$z-w=(a+bi)-(c+di)=(a-c)+(b-d)i.$$

For example,

$$(2+3i)+(4-i)=(2+4)+(3-1)i=6+2i.$$
$$(2+3i)-(4-i)=(2-4)+(3-(-1))i=-2+4i.$$
$$4(2+3i)-2(4-i)=8+12i-8+2i=14i.$$

The Product of Complex Numbers
Let $z=a+bi$ and $w=c+di$ be complex numbers. The **product** of z and w is defined by

$$\begin{aligned} zw &= (a+bi)(c+di)=a(c+di)+bi(c+di) \\ &= ac+adi+bci+bdi^2 \\ &= ac+adi+bci+bd(-1) \\ &= (ac-bd)+(ad+bc)i. \end{aligned}$$

For example,

$$\begin{aligned} (2-3i)(4+2i) &= 2\cdot 4+2\cdot 2i-3i\cdot 4-3i\cdot 2i \\ &= 8+4i-12i-6i^2 \\ &= 8+(4-12)i-6(-1)=14-8i. \end{aligned}$$

The Conjugate of a Complex Number
Let $z=a+bi$ be a complex number. The **conjugate** of z is denoted by $\bar{z}$ (read as "z bar") and is given by $\bar{z}=a-bi$. Note that to obtain the conjugate of a complex number, reverse the sign of the imaginary part. For example, if $z=2+3i$, $u=-4-2i$, $v=5i$ and $w=4$, then $\bar{z}=2-3i$, $\bar{u}=-4+2i$, $\bar{v}=-5i$, and $\bar{w}=4$.

Geometrically, the conjugate $\bar{z}$ of the complex number z is the reflection of z about the real (horizontal) axis, as illustrated in Figure 3.5.

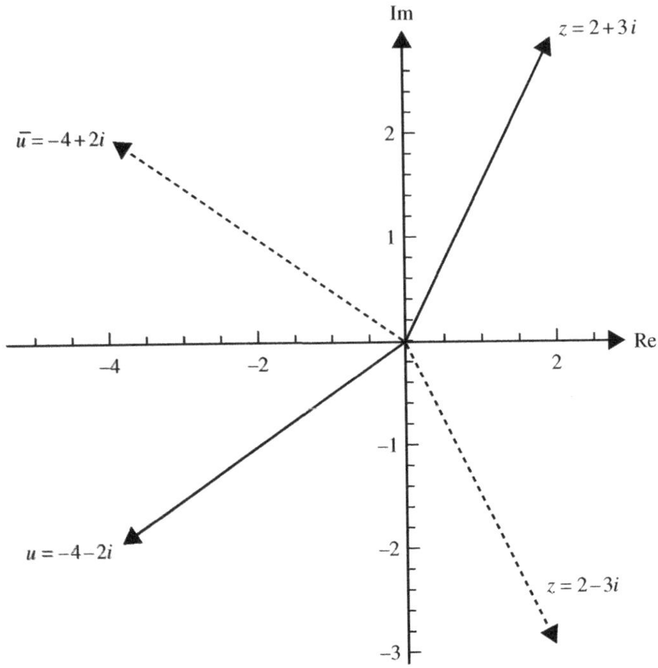

FIGURE 3.5. The conjugate of the complex number $z = 2 + 3i$ is the reflection of z about the real (horizontal) axis—that is, $\overline{z} = 2 - 3i$. The conjugate of $u = -4 - 2i$ is $u = -4 + 2i$, which is the reflection of u about the real axis.

The Absolute Value of a Complex Number

The **absolute value** (or **modulus**) of the complex number $z = a + bi$ is denoted by $|z|$ and is defined by $|z| = \sqrt{a^2 + b^2}$. Note that because a complex number $z = a + bi$ is a vector in the complex plane, the absolute value $|z|$ is the length of the vector, as shown in Figure 3.6.

For example, if $z = 3 + 4i$, $w = -2 + i$, $u = -2$, and $v = -3i$, then

$$|z| = \sqrt{3^2 + 4^2} = \sqrt{9+16} = \sqrt{25} = 5.$$

$$|w| = \sqrt{(-2)^2 + 1^2} = \sqrt{4+1} = \sqrt{5}.$$

$$|u| = \sqrt{(-2)^2 + 0^2} = \sqrt{4} = 2.$$

$$|v| = \sqrt{0^2 + (-3)^2} = \sqrt{9} = 3.$$

Division of Complex Numbers

Let $z = a + bi$ and $w = c + di$. The **division** $\dfrac{z}{w} = \dfrac{a+bi}{c+di}$ can be found by multiplying the numerator and the dominator by the conjugate of the dominator $\overline{w}$. For example,

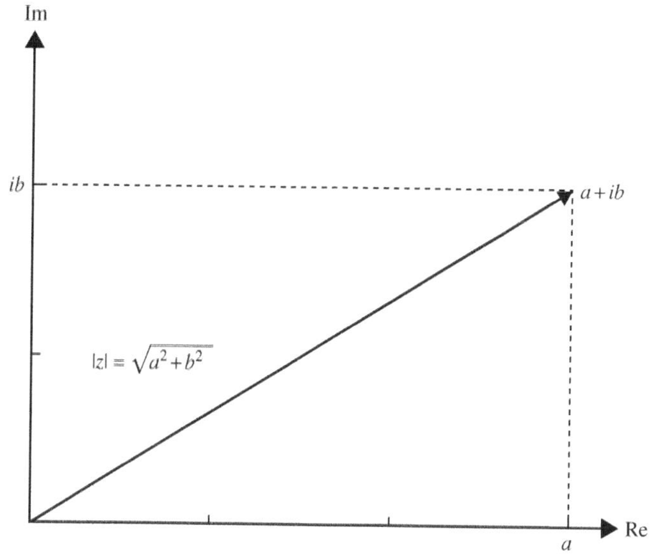

FIGURE 3.6. The absolute value (or modulus) |z| of the complex number $z=a+ib$ is $|z|=\sqrt{a^2+b^2}$.

$$\frac{2+3i}{4-2i}=\frac{2+3i}{4-2i}\cdot\frac{4+2i}{4+2i}$$

$$=\frac{8+4i+12i+6i^2}{16+8i-8i-4i^2}=\frac{8+16i+6(-1)}{16-4(-1)}=\frac{2+16i}{20}=\frac{2}{20}+\frac{16}{20}i=0.1+0.8i.$$

Properties of the Absolute Value

Let $z=a+bi$ and $w=c+di$ be complex numbers. The main properties of the absolute values are summarized in the following

$|z|=0$ if and only if $z=0$.

$$|z|=|\bar{z}|.$$

$$|zw|=|z||w|.$$

$$|z^n|=|z|^n.$$

$$\left|\frac{z}{w}\right|=\frac{|z|}{|w|},\quad w\neq 0.$$

$$|z+w|\leq|z|+|w|.$$

3.5.5. Complex Eigenvalues and Complex Eigenvectors

Knowing the basics of complex numbers and complex matrices we are equipped to study complex eigenvalues of matrices and the corresponding complex eigenvectors.

Example 3.28: Complex Eigenvalues and Complex Eigenvectors
Find the eigenvalues and the corresponding eigenvectors of the matrix

$$A = \begin{bmatrix} 1 & -3 \\ 3 & 1 \end{bmatrix}.$$

Solution
First, find the characteristic polynomial of the matrix A, $\det(A-\lambda I_2)$.

We have

$$A-\lambda I_2 = \begin{bmatrix} 1 & -3 \\ 3 & 1 \end{bmatrix} - \lambda \begin{bmatrix} 1 & 0 \\ 0 & 1 \end{bmatrix} = \begin{bmatrix} 1-\lambda & -3 \\ 3 & 1-\lambda \end{bmatrix}.$$

The characteristic polynomial of A is

$$\det(A-\lambda I_2) = \det\begin{bmatrix} 1-\lambda & -3 \\ 3 & 1-\lambda \end{bmatrix} = (1-\lambda)(1-\lambda)+9 = \lambda^2 - 2\lambda + 10.$$

The eigenvalues of A are the solutions of the equation $\det(A-\lambda I)=0$. We have $\lambda^2-2\lambda+10=0$. Recall that the solutions of a quadratic equation $ax^2+bx+c=0$ are $x = \dfrac{-b \pm \sqrt{b^2-4ac}}{2a}$. The solutions of the given equation are

$$\lambda = \frac{-(-2) \pm \sqrt{4-4\cdot 1\cdot 10}}{2\cdot 1} = \frac{2\pm\sqrt{-36}}{2} = \frac{2\pm 6i}{2} = 1\pm 3i.$$

Therefore, the eigenvalues of A are $\lambda_1 = 1+3i$ and $\lambda_2 = 1-3i$. Note that λ_2 is a conjugate of λ_1 and λ_1 is a conjugate of λ_2.

To find the eigenvectors corresponding to the eigenvalue $\lambda_1 = 1+3i$, solve the following matrix equation

$$(A-(1+3i)I)X = O \quad \text{for} \quad X = \begin{bmatrix} x_1 \\ x_2 \end{bmatrix}.$$

We have

$$\left(\begin{bmatrix} 1 & -3 \\ 3 & 1 \end{bmatrix} - (1+3i)\begin{bmatrix} 1 & 0 \\ 0 & 1 \end{bmatrix}\right)\begin{bmatrix} x_1 \\ x_2 \end{bmatrix} = \begin{bmatrix} 0 \\ 0 \end{bmatrix}.$$

$$\begin{bmatrix} 1-(1+3i) & -3 \\ 3 & 1-(1+3i) \end{bmatrix} \begin{bmatrix} x_1 \\ x_2 \end{bmatrix} = \begin{bmatrix} 0 \\ 0 \end{bmatrix}.$$

$$\begin{bmatrix} -3i & -3 \\ 3 & -3i \end{bmatrix} \begin{bmatrix} x_1 \\ x_2 \end{bmatrix} = \begin{bmatrix} 0 \\ 0 \end{bmatrix}.$$

$$\begin{bmatrix} -3ix_1 - 3x_2 \\ 3x_1 - 3ix_2 \end{bmatrix} = \begin{bmatrix} 0 \\ 0 \end{bmatrix}.$$

Thus

$$-3ix_1 - 3x_2 = 0.$$

$$3x_1 - 3ix_2 = 0.$$

To solve this homogenous system of two linear equations, we use the following augmented matrix:

$$\begin{bmatrix} -3i & -3 & 0 \\ 3 & -3i & 0 \end{bmatrix}.$$

The reduced echelon form of the augmented matrix is

$$\begin{bmatrix} 1 & -i & 0 \\ 0 & 0 & 0 \end{bmatrix}.$$

Therefore,

$$x_1 - ix_2 = 0.$$

Letting $x_1 = 1$, we get $x_2 = -i$. Therefore, the eigenvector corresponding to $\lambda_1 = 1 + 3i$ is $V_1 = \begin{bmatrix} 1 \\ -i \end{bmatrix}$. Similarly the eigenvector corresponding to $\lambda_2 = 1 - 3i$ is $V_2 = \begin{bmatrix} 1 \\ i \end{bmatrix}$.

3.5.6. Complex Eigenvalues and Eigenvectors with MATLAB

Let's use MATLAB to find the eigenvalues and the corresponding eigenvectors of the matrix

$$A = \begin{bmatrix} 1 & -3 \\ 3 & 1 \end{bmatrix}.$$

```
>> A=[1 -3; 3 1];
>> [V D]=eig(A)
V=
   0.7071                   0.7071
        0 -0.7071i               0+0.7071i
D=
   1.0000+3.0000i            0
        0                1.0000-3.0000i
```

This shows that $\lambda_1 = 1+3i$ and the corresponding eigenvector, in normalized form, is

$$\begin{bmatrix} 0.7071 \\ -0.7071i \end{bmatrix} = 0.7071\begin{bmatrix} 1 \\ -i \end{bmatrix}.$$

This means that the corresponding eigenvector to λ_1 is $V_1 = \begin{bmatrix} 1 \\ -i \end{bmatrix}$. Similarly, the corresponding eigenvector to $\lambda_2 = 1-3i$ is $V_2 = \begin{bmatrix} 1 \\ i \end{bmatrix}$.

Here's another way to obtain the same results using MATLAB:

```
>> A=[1 -3; 3 1];
>> ev=eig(A)
ev=
   1.0000+3.0000i
   1.0000-3.0000i
>> rref(A-ev(1)*eye(2))
ans=
   1.0000    0-1.0000i
        0            0
>> rref(A - ev(2)*eye(2))
ans=
   1.0000     0 +1.0000i
        0             0
```

We obtain the same values for V_1 and V_2 as earlier.

3.5.7. Exercises 3.5

1. Let $A = \begin{bmatrix} 0 & 1 \\ -4 & 4 \end{bmatrix}$.

A. Show that $X = \begin{bmatrix} 1 \\ 2 \end{bmatrix}$ is an eigenvector of A. Find the corresponding eigenvalue.

B. Show that $Y = \begin{bmatrix} 3 \\ 4 \end{bmatrix}$ is not an eigenvector of A.

2. Let $A = \begin{bmatrix} 1 & 6 \\ 5 & 2 \end{bmatrix}$. Show that −4 is an eigenvalue of A. Determine the eigenvectors corresponding to the eigenvalue −4.

In Exercises 3–7,

A. Find the eigenvalues and the corresponding eigenvectors of the matrix.

B. Use MATLAB to find the eigenvalues and the corresponding eigenvectors. Be sure that the eigenvalues and the eigenvectors are the same as in part A.

C. Give geometrical interpretations of the eigenvalues and the eigenvectors.

3. $A = \begin{bmatrix} 2 & 1 \\ -5 & -4 \end{bmatrix}$.

4. $A = \begin{bmatrix} 0.6 & 0.2 \\ 0.4 & 0.8 \end{bmatrix}$.

5. $A = \begin{bmatrix} 0 & 1 \\ -9 & 6 \end{bmatrix}$.

6. $A = \begin{bmatrix} 3 & 1 \\ 2 & 4 \end{bmatrix}$.

7. $A = \begin{bmatrix} -3 & 0 \\ 2 & 4 \end{bmatrix}$.

In Exercises 8–11, let $u=2+3i$, $v=5-2i$, and $z=-4+2i$.

8. Find $u+v$, $v-z$, $3v+5z$, and $-3u+vi$.

9. Find uv, zu, $\frac{u}{v}$, $\frac{z}{u}$, v^2, and v^3.

10. Find the conjugate vectors $\bar{u}$, $\bar{v}$, and $\bar{z}$, Represent the vectors and their conjugates graphically.

11. Find $|u|$, $|z|$, $|v^2|$, and $|uz|$.

In Exercises 12–14,

A. Find the eigenvalues and the corresponding eigenvectors of the matrix.

B. Use MATLAB to find the eigenvalues and the corresponding eigenvectors. Be sure that the eigenvalues and the eigenvectors are the same as in part A.

12. $A = \begin{bmatrix} 2 & 3 \\ -3 & 2 \end{bmatrix}$.

13. $A = \begin{bmatrix} 3 & -6 \\ 6 & 3 \end{bmatrix}$.

14. $A = \begin{bmatrix} 1.08 & -0.2 \\ 0.1 & 0.84 \end{bmatrix}$.

15. Consider a species that lives for 2 years. Group the population in age classes [0, 1) and [1, 2]. Assume that the species has the following characteristics:

    i. The survival rates of the age classes [0, 1) and [1, 2] are 0.5 and 0, respectively.
    ii. The average number of offspring produced by the [0, 1) and [1, 2] age classes are 1 and 6, respectively.

    Let a_n and b_n be the population of the species in the age classes [0, 1) and [1, 2] after n years, respectively.

    A. Model this situation by a system of difference equations and by a single matrix difference equation.
    B. Use MATLAB to find the eigenvalues and the corresponding eigenvectors.
    C. Interpret the meaning of the eigenvalues and the eigenvectors in the context of this model.
    D. For the eigenvalues that have biological meaning, use MATLAB to determine a_n and b_n for large values of n and for different initial values, including initial value vectors equal to the eigenvectors corresponding to the positive eigenvalues. Explain your conclu sions.

16. Consider a species that lives for 3 years. Group the population in age classes [0, 1), [1, 2), and [2, 3]. Assume that the species has the following characteristics:

    i. The survival rate of the age classes [0, 1), [1, 2), and [2, 3] are 0.6, 0.4, and 0, respectively.
    ii. The average number of offspring produced by the [0, 1), [1, 2), and [2, 3] age classes are 0, 5, and 25/3, respectively.

    Let a_n, b_n, and c_n be the population of the species in the age classes [0, 1), [1, 2), and [2, 3] after n years, respectively.

    A. Model this situation by a system of difference equations and by a single matrix difference equation.
    B. Use MATLAB to find the eigenvalues and the corresponding eigenvectors.
    C. Interpret the meaning of the eigenvalues and the eigenvectors in the context of this model.
    D. For the eigenvalues that have biological meaning, use MATLAB to determine a_n, b_n, and c_n for large values of n and for different initial values, including initial value vectors equal to the eigenvectors corresponding to the positive eigenvalues. Explain your conclusions.

3.6. EIGENVALUES AND STABILITY OF LINEAR MODELS

In this section we investigate the use of eigenvalues to determine the long-term behavior of a system of linear equations.

Recall that the general solution of the one-dimensional linear homogenous difference equation $x_{n+1} = ax_n$ is $x_n = a^n x_0$, where a is a real number. The value of a determines the behavior of the system. If $|a| < 1$, the system approaches a unique fixed point 0, and if $|a| > 1$, the system does not converge to a fixed point but gets larger and larger without bound.

Consider a system of linear difference equations represented by the following matrix equation:

$$X_{n+1} = AX_n,$$

where A is $n \times n$ matrix and X_n is an $n \times 1$ vector. Knowing the initial vector X_0, we have

$$X_1 = AX_0.$$

$$X_2 = AX_1 = A(AX_0) = A^2 X_0.$$

$$\dots\dots$$

$$X_n = A^n X_0.$$

This equation gives us the exact form of X_n but it does not provide us with the general behavior of the system. We would like to use matrix A to inform us about the dynamics of the system.

3.6.1. Investigation 1

Consider the population movement model between a city and its surrounding suburbs studied in Model 3.6. Recall that the system is represented by the following matrix equation:

$$X_{n+1} = TX_n,$$

where

$$X_n = \begin{bmatrix} c_n \\ s_n \end{bmatrix}, \quad T = \begin{bmatrix} 0.92 & 0.02 \\ 0.08 & 0.98 \end{bmatrix}.$$

Using MATLAB to calculate the eigenvalues and the corresponding eigenvectors of transition matrix T, we get two eigenvalues, $\lambda_1 = 1$ and $\lambda_2 = 0.9$, and the corresponding eigenvectors are $t\begin{bmatrix} 0.25 \\ 1 \end{bmatrix}$ and $s\begin{bmatrix} -1 \\ 1 \end{bmatrix}$, respectively, where t and s are parameters.

Assume that the total population in the city and suburb is 10 million. Let's use MATLAB to investigate the long-term behavior of the system, where components of X_0 are in millions.

$$X_0 = \begin{bmatrix} 4 \\ 6 \end{bmatrix}.$$

$$T^{50} X_0 = \begin{bmatrix} 2.0103 \\ 7.9897 \end{bmatrix}.$$

$$T^{100} X_0 = \begin{bmatrix} 2.0001 \\ 7.9999 \end{bmatrix}.$$

$$T^{101} X_0 = \begin{bmatrix} 2.0000 \\ 8.0000 \end{bmatrix}.$$

$$T^{102} X_0 = \begin{bmatrix} 2.0000 \\ 8.0000 \end{bmatrix}.$$

These indicate that the vector $\begin{bmatrix} 2 \\ 8 \end{bmatrix}$ is an equilibrium vector (steady state vector). Similarly, if we use any other initial vector, the system will approach vector $\begin{bmatrix} 2 \\ 8 \end{bmatrix}$. It can be concluded that vector $\begin{bmatrix} 2 \\ 8 \end{bmatrix}$ is the equilibrium vector of the system. Letting $V_1 = \begin{bmatrix} 2 \\ 8 \end{bmatrix}$, we have $TV_1 = V_1$, which means that V_1 is an eigenvector corresponding to the eigenvalue $\lambda = 1$. Note that the eigenvector V_1 is the same as the eigenvector $t\begin{bmatrix} 0.25 \\ 1 \end{bmatrix}$ with $t=8$. Note also that the eigenvector $V_2 = s\begin{bmatrix} -1 \\ 1 \end{bmatrix}$ with any value of s is not an equilibrium vector because $TV_2 = 0.9V_2$.

Consider this equation:

$$X_{n+1} = AX_n, \tag{3.6}$$

and let V be an eigenvector corresponding to the eigenvalue λ. Letting $X_0 = V$ we have

$$X_1 = AX_0 = AV = \lambda V.$$

$$X_2 = AX_1 = A(\lambda V) = \lambda AV = \lambda(\lambda V) = \lambda^2 V.$$

$$X_3 = AX_2 = A(\lambda^2 V) = \lambda^2 AV = \lambda^2 (\lambda V) = \lambda^3 V.$$

Similarly,

$$X_n = \lambda^n V,$$

which provides a simple formula for calculating X_n for any value of n.

Because there is no negative population, it seems that the eigenvector $s\begin{bmatrix}-1\\1\end{bmatrix}$ corresponding to the eigenvalue 0.9 is not relevant in studying this population model. However, this is not the case. We will see how any initial population $X_0 = \begin{bmatrix}x_0\\y_0\end{bmatrix}$ can be expressed as a linear combination of the two eigenvectors—that is,

$$X_0 = c_1\begin{bmatrix}1\\4\end{bmatrix} + c_2\begin{bmatrix}-1\\1\end{bmatrix}.$$

$$\begin{bmatrix}x_0\\y_0\end{bmatrix} = \begin{bmatrix}1 & -1\\4 & 1\end{bmatrix}\begin{bmatrix}c_1\\c_2\end{bmatrix}.$$

where c_1 and c_2 are scalars (constants).

For instance, if $X_0 = \begin{bmatrix}3\\7\end{bmatrix}$, we have

$$\begin{bmatrix}c_1\\c_2\end{bmatrix} = \begin{bmatrix}1 & -1\\4 & 1\end{bmatrix}^{-1}\begin{bmatrix}3\\7\end{bmatrix} = \begin{bmatrix}2\\-1\end{bmatrix}.$$

Therefore

$$\begin{bmatrix}3\\7\end{bmatrix} = 2\begin{bmatrix}1\\4\end{bmatrix} - \begin{bmatrix}-1\\1\end{bmatrix}.$$

In general, if λ_1 and λ_2 are real distinct eigenvalues, and V_1 and V_2 the corresponding eigenvectors of the 2×2 matrix A of equation 3.6, then the initial vector X_0 can be expressed as

$$X_0 = c_1V_1 + c_2V_2, \tag{3.7}$$

where c_1 and c_2 are scalars. We have

$$X_1 = AX_0 = A(c_1V_1 + c_2V_2) = Ac_1V_1 + Ac_2V_2 = c_1AV_1 + c_2AV_2 = c_1\lambda_1V_1 + c_2\lambda_2V_2.$$

$$X_2 = AX_1 = A(c_1\lambda_1V_1 + c_2\lambda_2V_2) = Ac_1\lambda_1V_1 + Ac_2\lambda_2V_2 = c_1\lambda_1AV_1 + c_2\lambda_2AV_2 = c_1\lambda_1^2V_1 + c_2\lambda_2^2V_2.$$

Similarly,

$$X_k = c_1\lambda_1^kV_1 + c_2\lambda_2^kV_2. \tag{3.8}$$

The long-term behavior of the system depends on the eigenvalues λ_1 and λ_2. For example, in this model we have $c_1=2$, $c_2=-1$, $\lambda_1=1$, $\lambda_2=0.9$, $V_1=\begin{bmatrix}1\\4\end{bmatrix}$, and $V_2=\begin{bmatrix}-1\\1\end{bmatrix}$. Therefore, $X_k \to \begin{bmatrix}2\\8\end{bmatrix}$ as k becomes large.

This result confirms the conclusion that the steady state vector is $\begin{bmatrix}2\\8\end{bmatrix}$.

We will generalize the results to systems represented by the following matrix equation:

$$X_{n+1} = AX_n, \tag{3.9}$$

Let's summarize the long-term behavior of a system represented by the following matrix equation:

$$X_{n+1} = AX_n,$$

where A is a 2×2 matrix. Assume that A has two real distinct eigenvalues, λ_1 and λ_2. The corresponding eigenvectors are V_1 and V_2. The general solution is

$$X_k = c_1\lambda_1^k V_1 + c_2\lambda_2^k V_2.$$

The system's long-term behavior is determined based on the values of the eigenvalues as follows:

- If $|\lambda_1|<1$ and $|\lambda_2|<1$, then λ_1^k and λ_2^k both go to zero as k becomes large, and consequently $X_k \to \begin{bmatrix}0\\0\end{bmatrix}$ as $k\to\infty$. Note that this does not depend on the values of c_1 and c_2. Therefore, the zero vector is a **stable equilibrium solution**.
- If $\lambda_1=1$ and $|\lambda_2|<1$, then $\lambda_2^k \to 0$. Consequently, $X_k\to c_1V_1$. Similarly, if $|\lambda_1|<1$ and $\lambda_2=1$, then $\lambda_1^k \to 0$ and $X_k\to c_2V_2$. Note that in the first case c_1V_1 is an equilibrium solution and in the second case c_2V_2 is an equilibrium solution. In both cases, the zero vector is an unstable equilibrium solution because the zero vector is never approached.
- If $\lambda_1=-1$ and $|\lambda_2|<1$, then $\lambda_2^k \to 0$ and $X_k\to c_1(-1)^kV_1$, which oscillates between c_1V_1 and $-c_1V_1$. Therefore, the zero vector is an unstable equilibrium solution, unless $c_1=0$. Similarly, if $|\lambda_1|<1$ and $\lambda_2=-1$, then $X_k\to c_2(-1)^kV_2$ and if $|\lambda_1|<1$ and $|\lambda_2|>1$, then $\lambda_2^k \to\infty$. Therefore, $X_k\to\infty$. Similarly, if $|\lambda_1|>1$ and $|\lambda_2|<1$, then $\lambda_1^k \to\infty$ and $X_k\to\infty$
- If $|\lambda_1|>1$ and $|\lambda_2|>1$, then $X_k\to\infty$.

where A is $n \times n$ matrix. Assume that A has n real distinct eigenvalues, $\lambda_1, \lambda_2, \ldots, \lambda_n$, and the corresponding eigenvectors are $V_1, V_2, \ldots, V_n$.

An eigenvalue λ_d of A is called a **dominant eigenvalue** if

$$|\lambda_d| \geq |\lambda_i|, \qquad i = 1, 2, \ldots, \quad \text{and} \quad i \neq d,$$

and the eigenvector V_d corresponding to λ_d is called a **dominant eigenvector** if

$$|\lambda_d| > |\lambda_i|, \qquad i = 1, 2, \ldots, n \quad \text{and} \quad i \neq d.$$

Then the λ_d is called a **strictly dominant eigenvalue**.

An initial vector, X_0, can be represented as

$$X_0 = c_1 V_1 + c_2 V_2 + \ldots + c_n V_n,$$

where $c_1, c_2, \ldots, c_n$ are scalars (constants). It can be easily shown that the solution of equation 3.9 is

$$X_k = c_1 \lambda_1^k V_1 + c_2 \lambda_2^k V_2 + \ldots + c_n \lambda_n^k V_n. \tag{3.10}$$

Because we assumed that there are n real distinct eigenvalues, there is a strictly dominant eigenvalue. For simplicity we number the strictly dominant eigenvalue as λ_1 and the corresponding (dominant) eigenvector as V_1. We have

$$X_k = \lambda_1^k \left(c_1 V_1 + c_2 \left(\frac{\lambda_2}{\lambda_1} \right)^k V_2 + \ldots + c_n \left(\frac{\lambda_n}{\lambda_1} \right)^k V_n \right).$$

Because λ_1 is the strictly dominant eigenvalue, $\left| \frac{\lambda_i}{\lambda_1} \right| < 1$ for $i = 2, 3, \ldots, n$ and $\left| \left(\frac{\lambda_i}{\lambda_1} \right)^k \right| \to 0$ as k become very large. Consequently,

$$X_k \to \lambda_1^k c_1 V_1 \tag{3.11}$$

The long-term behavior of the system depends on the value of the dominant eigenvalue λ_1, and we have the following three cases:

Case 1: $\lambda_1 < 1$. If $\lambda_1 < 1$, then X_k exponentially decays. That is, $X_k \to 0$ as k becomes very large.

Case 2: $\lambda_1 > 1$. If $\lambda_1 > 1$, then X_k exponentially grows. That is, X_k becomes unbounded large as k becomes very large.

Case 3: $\lambda_1 = 1$. If $\lambda_1 = 1$, then $X_k \to c_1 V_1$.

3.6.2. Repeated Eigenvalues

Consider a system of two linear difference equations represented by the following matrix equation

$$X_{n+1} = AX_n,$$

where A is a 2×2 matrix and $X_0 = \begin{bmatrix} x_0 \\ y_0 \end{bmatrix}$.

So far we considered examples in which a 2×2 matrix A had two real distinct eigenvalues, λ_1 and λ_2, with the corresponding eigenvectors V_1 and V_2, respectively. In this case, any initial vector, X_0 may be expressed as a linear combination of V_1 and V_2—that is,

$$X_0 = c_1V_1 + c_2V_2,$$

where c_1 and c_2 are scalars. Moreover, the solution of equation 3.6, X_k, is given in the form

$$X_k = c_1\lambda_1^kV_1 + c_2\lambda_2^kV_2.$$

This equation determines the long-term behavior of the system.

We are interested in finding the general solution of equation 3.6 when the matrix A has one repeated real eigenvalue, λ, and one corresponding eigenvector, V_1. Similar to the case of two real distinct eigenvalues, we expect that the term $c_1\lambda^kV_1$ will be a part of the formula for X_k. Because there is no second eigenvalue, λ_2, and its corresponding eigenvector, we will try the conjecture that

$$X_k = k\lambda^kV_1 + \lambda^kV_2,$$

where V_2 is a vector to be determined. We have $X_{n+1} = (n+1)\lambda^{n+1}V_1 + \lambda^{n+1}V_2$ and $X_n = n\lambda^nV_1 + \lambda^nV_2$. By substitution in equation 3.6, we get

$$(n+1)\lambda^{n+1}V_1 + \lambda^{n+1}V_2 = A(n\lambda^nV_1 + \lambda^nV_2).$$

$$(n+1)\lambda^{n+1}V_1 + \lambda^{n+1}V_2 = n\lambda^nAV_1 + \lambda^nAV_2.$$

Because $AV_1 = \lambda V_1$, this equation simplifies to

$$\lambda V_1 + \lambda V_2 = AV_2$$

or

$$(A - \lambda I)V_2 = \lambda V_1. \tag{3.12}$$

Therefore, $X_k = k\lambda^k V_1 + \lambda^k V_2$ is a solution of equation 3.6 if equation 3.12 is satisfied. Consequently, the general solution of equation 3.6 is

$$X_k = c_1\lambda^k V_1 + c_2(k\lambda^k V_1 + \lambda^k V_2),$$

where V_2 is determined from equation 3.12.

Example 3.29

Find the general solution of the system $X_{n+1} = AX_n$, where $A = \begin{bmatrix} 1 & 1 \\ -1 & 3 \end{bmatrix}$.

Solution

Matrix A has one repeated eigenvalue, $\lambda = 2$, and the corresponding eigenvector, $V_1 = \begin{bmatrix} 1 \\ 1 \end{bmatrix}$ (see Example 3.27). Use equation 3.12 to determine the vector $V_2 = \begin{bmatrix} x \\ y \end{bmatrix}$. We have

$$\left(\begin{bmatrix} 1 & 1 \\ -1 & 3 \end{bmatrix} - 2\begin{bmatrix} 1 & 0 \\ 0 & 1 \end{bmatrix}\right)\begin{bmatrix} x \\ y \end{bmatrix} = 2\begin{bmatrix} 1 \\ 1 \end{bmatrix}.$$

$$\begin{bmatrix} -1 & 1 \\ -1 & 1 \end{bmatrix}\begin{bmatrix} x \\ y \end{bmatrix} = \begin{bmatrix} 2 \\ 2 \end{bmatrix}.$$

The reduced echelon form of the augmented matrix is

$$\begin{bmatrix} 1 & -1 & -2 \\ 0 & 0 & 0 \end{bmatrix}.$$

This implies that $\begin{bmatrix} x \\ y \end{bmatrix} = \begin{bmatrix} r-2 \\ r \end{bmatrix}$, where r is a parameter. Letting $r = 2$, we obtain $V_2 = \begin{bmatrix} 0 \\ 2 \end{bmatrix}$. Consequently, the general solution of equation 3.6 is

$$X_k = c_1 2^k \begin{bmatrix} 1 \\ 1 \end{bmatrix} + c_2\left(k2^k \begin{bmatrix} 1 \\ 1 \end{bmatrix} + 2^k \begin{bmatrix} 0 \\ 2 \end{bmatrix}\right),$$

where c_1 and c_2 are constants.

3.6.3. Complex Eigenvalues

We are interested in the long-term behavior of a system having complex eigenvalues and complex eigenvectors.

Consider the following system:

$$X_{k+1} = AX_k, \tag{3.13}$$

where A is an $n \times n$ matrix. Assume that matrix A has n eigenvalues, $\lambda_1, \lambda_2, \ldots, \lambda_n$, and the corresponding eigenvectors are $V_1, V_2, \ldots, V_n$. We assume that some of these eigenvalues are complex. Note that the complex eigenvalues are conjugate.

We will see that the results about the long-term behavior of a system with n real distinct eigenvalues are applicable to the current case in which some of the eigenvalues are complex. Assume that λ_1 is a strictly dominant eigenvalue—that is, $|\lambda_1| > |\lambda_j|, j = 2, 3, \ldots, n$. From the properties of the absolute value of complex numbers discussed in the last section, we have $\left|\frac{\lambda_j}{\lambda_1}\right| < 1$. Consequently, $\left|\frac{\lambda_j}{\lambda_1}\right|^k \to 0$ for large values of k. This is equivalent to $\left|\left(\frac{\lambda_j}{\lambda_1}\right)^k\right| \to 0$; it can be concluded that $\left(\frac{\lambda_j}{\lambda_1}\right)^k \to 0$ and $X_k \to \lambda_1^k c_1 V_1$.

Example 3.30

Determine the long term-behavior of a system represented by the following difference equation:

$$X_{n+1} = AX_n, \tag{3.14}$$

where $A = \begin{bmatrix} 0 & 0 & 40 \\ 0.03 & 0 & 0 \\ 0 & 0.25 & 0.54 \end{bmatrix}$.

Solution

Using MATLAB, matrix A has three eigenvalues: $\lambda_1 = 0.9057$, $\lambda_2 = -0.1829 + 0.5457i$, and $\lambda_3 = -0.1829 - 0.5457i$. The corresponding eigenvectors are

$$V_1 = \begin{bmatrix} 0.9992 \\ 0.0331 \\ 0.0226 \end{bmatrix}.$$

$$V_2 = \begin{bmatrix} 0.9985 \\ -0.0165 - 0.0494i \\ -0.0046 + 0.0136i \end{bmatrix}.$$

$$V_3 = \begin{bmatrix} 0.9985 \\ -0.0165+0.0494i \\ -0.0046-0.0136i \end{bmatrix}.$$

Because λ_1 is the dominant eigenvalue, the solution of equation 3.14 is

$$X_k = \lambda_1^k \left(c_1 V_1 + c_2 \left(\frac{\lambda_2}{\lambda_1} \right)^k + c_3 \left(\frac{\lambda_3}{\lambda_1} \right)^k \right).$$

Because $|\lambda_1| = 0.9057 < 1$, we have

$$X_k \to \lambda_1^k c_1 V_1 = (0.9057)^k c_1 \begin{bmatrix} 0.9992 \\ 0.0331 \\ 0.0226 \end{bmatrix} \to 0.$$

3.6.4. Exercises 3.6

In Exercises 1–12, each system is represented by a matrix equation in the form $X_{n+1} = TX_n$.

A. Find the eigenvalues and the corresponding eigenvector of matrix T.

B. Use MATLAB to check your answer in part A.

C. Write the form of the general solution and predict the long term behavior of the system.

1. $T = \begin{bmatrix} 3 & 1 \\ 2 & 4 \end{bmatrix}.$

2. $T = \begin{bmatrix} -3 & 0 \\ 2 & 4 \end{bmatrix}.$

3. $T = \begin{bmatrix} 2 & 1 \\ -5 & -4 \end{bmatrix}.$

4. $T = \begin{bmatrix} 0.6 & 0.2 \\ 0.4 & 0.8 \end{bmatrix}.$

5. $T = \begin{bmatrix} 0 & 1 \\ -9 & 6 \end{bmatrix}.$

6. $T = \begin{bmatrix} 1 & 2 \\ 0.3 & 0 \end{bmatrix}$.

7. $T = \begin{bmatrix} 2 & 3 \\ -3 & 2 \end{bmatrix}$.

8. $T = \begin{bmatrix} 1.08 & -0.2 \\ 0.1 & 0.84 \end{bmatrix}$.

9. Consider a species that lives for 2 years. Group the population in age classes [0, 1) and [1, 2]. Assume that the species has the following characteristics:

   i. The survival rates of the age classes [0, 1) and [1, 2] are 0.5 and 0, respectively.
   ii. The average number of offspring produced by the [0, 1) and [1, 2] age classes are 1 and 6, respectively.

   Let a_n and b_n be the population of the species in the age classes [0, 1) and [1, 2] after n years, respectively.

   A. Model this situation by a system of difference equations and by a single matrix difference equation.
   B. Use MATLAB to find the eigenvalues and the corresponding eigenvectors.
   C. Interpret the meaning of the eigenvalues and the eigenvectors in the context of this model.
   D. Determine the long-term behavior of the system.

10. Consider a species that lives for 3 years. Group the population in age classes [0, 1), [1, 2), and [2, 3]. Assume that the species has the following characteristics:

   i. The survival rate of the age classes [0, 1), [1, 2), and [2, 3] are 0.6, 0.4, and 0, respectively.
   ii. The average number of offspring produced by the [0, 1), [1, 2), and [2, 3] age classes are 0, 5, and 25/3, respectively.

   Let a_n, b_n, and c_n be the population of the species in the age classes [0, 1), [1, 2), and [2, 3] after n years, respectively.

   A. Model this situation by a single matrix difference equation.
   B. Use MATLAB to find the eigenvalues and the corresponding eigenvectors.
   C. Use the eigenvalues methods discussed in Section 3.6 to determine the long-term behaviour of the system.

CHAPTER 4

MODELING WITH SYSTEMS OF LINEAR DIFFERENCE EQUATIONS

4.1. MODELING WITH MARKOV CHAINS

In this section we discuss a special class of finite stochastic processes. We consider a sequence of finite experiments in which the outcomes and associated probabilities at each stage depend only on the outcomes and associated probabilities of the immediately preceding stage. Such a process is called a **Markov process** or a **Markov chain**. The Markov chains (processes) are named after the Russian mathematician Andrey Andreyevich Markov (1856–1922). Let's investigate the following population movement model.

4.1.1. A Population Movement Model

We consider a simple model of population movement between a certain city and its surrounding suburbs. For simplicity, we assume the following:

i. The people who move from the city go to the suburbs, and the people who move from the suburbs go to the city,
ii. We assume that during a year the total population in the city and its surrounding suburbs is fixed—that is we ignore other factors such as births and deaths.

Under these two assumptions the total population in the city and the suburbs is the same every year. Assume that the demographic studies showed that during 2000, 6% of the

Explorations of Mathematical Models in Biology with MATLAB®, First Edition. Mazen Shahin.

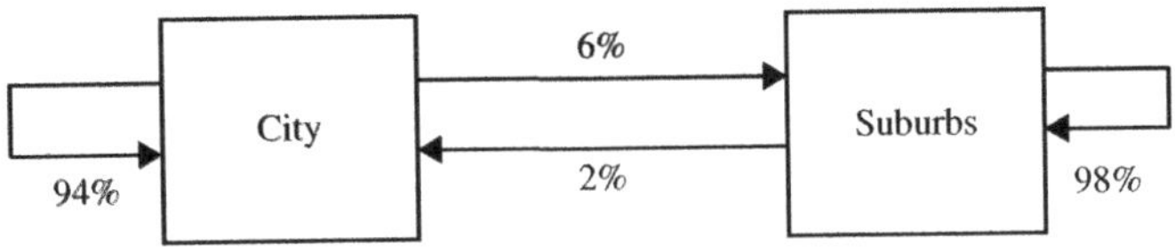

FIGURE 4.1. A diagram of the population movement probabilities between a city and its surrounding suburbs.

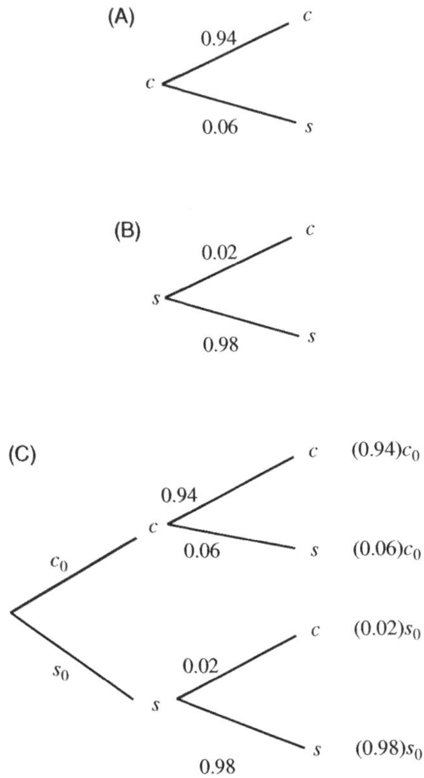

FIGURE 4.2. A. The population movement probabilities from a city in 1 year. **B**. The population movement probabilities from the suburbs in 1 year. **C**. A tree diagram showing the movement probabilities between a city and its surrounding suburbs after 1 year.

city population move to suburbs and 2% of the suburb population move to the city. This means that 94% of the city population stays in the city and 98% of the suburb population stays in the suburb.

Figure 4.1 illustrates these population movement probabilities.

This is a two-state Markov process. Every year (stage) there are two states: "live in city" and "live in suburbs." The probabilities in any stage (year) depend only on the previous stage (year). One way to model this situation is to use the probability tree. Figure 4.2A illustrates the assumption that 94% of the people who live in the city will stay in the city and 6% will move to the suburbs, and panel B illustrates the movement assumptions about the people who live in the suburbs.

Let's consider 2010 to be the initial year. Now assume that the initial populations of the city and the suburbs are c_0 and s_0, respectively. The probability tree after one year is shown in Figure 4.2C. Because there are two branches for c with probabilities $0.94c_0$ and $0.02s_0$, the city population after one year, c_1, is given by

$$c_1 = 0.94c_0 + 0.02s_0.$$

Similarly the suburbs population after one year, s_1, is given by

$$s_1 = 0.06c_0 + 0.98s_0.$$

In general, this Markov process is modeled by the following system of two first-order linear difference equations:

$$c_{n+1} = 0.94c_n + 0.02s_n \tag{4.1}$$

$$s_{n+1} = 0.06c_n + 0.98s_n, \tag{4.2}$$

where c_n and s_n are the city and suburbs populations after n years (from 2010).

Exploration 4.1: A Population Movement Model

Assume that the initial populations (at 2010) of the city and the suburbs are 3 million and 7 million, respectively, and assume that the migration percentages remain constant. Because the total number of people who live in the city and the suburbs remains constant, we prefer to use the proportions of the city and suburbs populations for c_n and s_n instead of the actual values of the populations. In this case $c_0=0.30$ and $s_0=0.70$. Note that $c_n+s_n=1$ for every value of n, $n=0, 1, 2, \ldots$

Using equations 4.1 and 4.2, we have

$$c_1 = 0.94(0.30) + 0.02(0.70) = 0.2960$$

and

$$s_1 = 0.06(0.30) + 0.98(0.70) = 0.7040.$$

Similarly,

$$c_2 = 0.94(0.2960) + 0.02(0.7040) = 0.2923$$

and

$$s_2 = 0.06(0.2960) + 0.98(0.7040) = 0.7077.$$

To conduct investigations and to study the long-term behavior of such a Markov chain we develop a MATLAB function to iterate the general form for a two-state Markov chain in the form

$$x_{n+1} = ax_n + by_n \tag{4.3}$$

and

$$y_{n+1} = cx_n + dy_n, \tag{4.4}$$

where a, b, c, and d are constant probabilities for transitions from one stage to the following stage.

```
function [T, X, Y] = Markov_2S (a, b, c, d, x0, y0, n)
% Two-state Markov process
%   x(n+1)=ax(n)+by(n)
%   y(n+1)=cx(n)+dy(n)
% Function parameters (input):
% Equations' constants a, b, c, and d
% Initial values x(0) and y(0) in proportions to the total x(0)+y(0)
% n=number of time periods
% Function output:
% T=sequence of time
% X=sequence of x(n)
% Y=sequence of y(n)

T=0:n;
X(1)=x0;
Y(1)=y0;
for j=1:n
  X(j+1)=a*X(j)+b*Y(j);
  Y(j+1)=c*X(j)+d*Y(j);
end;
plot(T, X, 'k.', T, Y, 'k*');
xlabel ('Time n in years');
ylabel ('City population x(n) & Suburb population y(n)');
legend ('City population', 'Suburb Population');
return;
```

To iterate Markov chain 4.1 and 4.2 with $c_0=0.30$, $s_0=0.70$, and $n=100$ years, we call the function **Markov_2S** by writing the following code:

```
>> [T, X, Y]=Markov_2S (0.94, 0.02, 0.06, 0.98, 0.30,
0.70, 100);
```

The graph is shown in Figure 4.3.

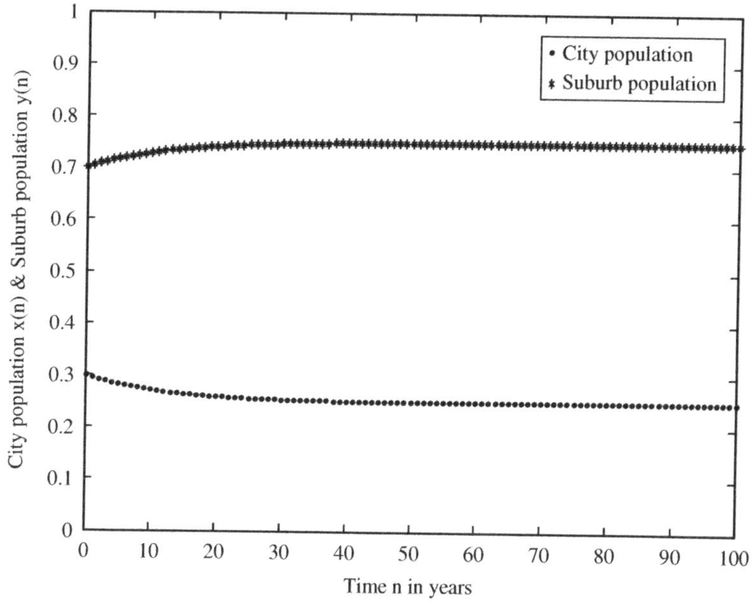

FIGURE 4.3. Graphs of a city and its surrounding suburbs populations, c_n and s_n, vs. time, n, in years, $n=0, 1, \ldots, 100$ of the two-state Markov chain $c_{n+1}=0.94c_n+0.02s_n$ and $s_{n+1}=0.06c_n+0.98s_n$, with the initial values $c_0=0.30$ and $s_n=0.70$.

Note that if this command is not followed by a semicolon, MATLAB will display all values of the vectors T, X, and Y. A simple way to display a portion of the these arrays, for example the first eight values, is with the following command:

```
>> [T(1:8); X(1:8); Y(1:8)]
```

With the initial conditions, $x(0) = 0.30$ and $y(0) = 0.70$, we have

$$X(83) = 0.2501$$
$$X(84) = 0.2500$$
$$X(85) = 0.2500,\ldots$$

and

$$Y(83) = 0.7499$$
$$Y(84) = 0.7500$$
$$Y(85) = 0.7500,\ldots.$$

These numerical values and the graph show that for large values of n, c_n reaches 0.25 and stays without change at that value, and s_n reaches 0.75 and stays without change at that value. The values $c_n = 0.25$ and $s_n = 0.75$ are called the **equilibrium values** or the **steady state values**.

Let's find the equilibrium values of systems 4.1 and 4.2 algebraically. Let c_e and s_e be the equilibrium values of c_n and s_n, respectively, at the equilibrium values $c_{n+1}=c_n=c_e$ and $s_{n+1}=s_n=s_e$. Substituting these values in the equations, we get

$$c_e = 0.94c_e + 0.02s_e.$$

$$s_e = 0.06c_e + 0.98s_e.$$

Simplify the first equation to get

$$0.06c_e - 0.02s_e = 0.$$

Note that the simplification of the second equation would lead also to the same equation, $0.06c_e - 0.02s_e = 0$. Because $c_e + s_e = 1$, we solve the two equations:

$$0.06c_e - 0.02s_e = 0.$$

$$c_e + s_e = 1.$$

We obtain $c_e = 0.25$ and $s_e = 0.75$.

Does the equilibrium values depend on the initial values c_0 and s_0? It is clear that the calculation of c_e and s_e did not include c_0 and s_0. Consequently, the equilibrium values do not depend on initial values. Figure 4.4 shows the graphs of c_n and s_n of systems 4.1 and 4.2 with the different initial values $c_0=0.40$ and $s_0=0.60$. It shows

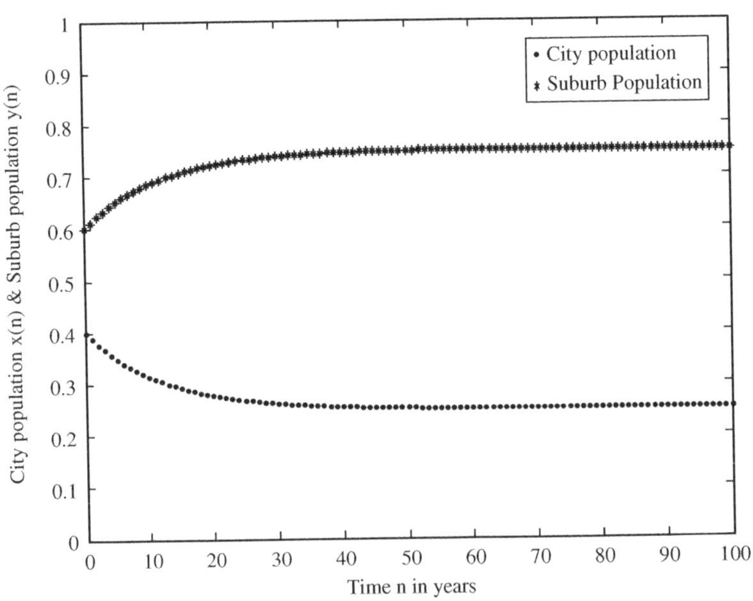

FIGURE 4.4. Graphs of a city and its surrounding suburbs populations, c_n and s_n, vs. time, n, in years, $n=0, 1, \ldots, 100$ of the two-state Markov chain $c_{n+1}=0.94c_n+0.02s_n$ and $s_{n+1}=0.06c_n+0.98s_n$, with the initial values $c_0=0.40$ and $s_n=0.60$.

that the equilibrium values are the same, $c_e=0.25$ and $s_e=0.75$. The only difference is that in this case c_n and s_n reach the equilibrium values c_e and s_e at $n=98$, that is,

$$c_{98} = c_{99} = c_e = 0.25 \quad \text{and} \quad s_{98} = s_{99} = s_e = 0.75.$$

In summary, the initial values do not affect the equilibrium values, but they affect the time at which the equilibrium values have been reached.

4.1.2. Matrix Representation of Markov Chains

We will model Markov chains by matrices and will demonstrate the advantages of the matrix representation, in particular, in studying the long-term behavior of Markov chains.

Exploration 4.2: The Population Movement Model Revisited

Let's reconsider the population movement between a city and its surrounding suburbs modeled by the difference equations 4.1 and 4.2 with the initial values $c_0=0.30$ and $s_0=0.70$,

$$c_{n+1} = 0.94c_n + 0.02s_n \tag{4.1}$$

$$s_{n+1} = 0.06c_n + 0.98s_n \tag{4.2}$$

This system can be represented by the following first-order linear matrix difference equation:

$$X_{n+1} = TX_n, \tag{4.5}$$

where $X_n = \begin{bmatrix} c_n \\ s_n \end{bmatrix}$ and $T = \begin{bmatrix} 0.94 & 0.02 \\ 0.06 & 0.98 \end{bmatrix}$. Note that

$$X_0 = \begin{bmatrix} c_0 \\ s_0 \end{bmatrix} = \begin{bmatrix} 0.30 \\ 0.70 \end{bmatrix}. \tag{4.6}$$

The column matrix (vector), X_n, is called the **distribution vector** or **state vector**, which represents the city and suburbs populations after n years. Note that the entries of matrix T in equation 4.6 represent the probabilities of moving from one state to another as follows,

$$T = \begin{array}{c} \text{From} \\ \begin{array}{cc} \text{City} & \text{Suburb} \end{array} \\ \left[\begin{array}{cc} 0.94 & 0.02 \\ 0.06 & 0.98 \end{array}\right] \end{array} \begin{array}{l} \text{City} \\ \text{Suburb} \end{array} \quad \text{To.}$$

Matrix T is called the **matrix of transition probabilities**, or the **transition matrix**. Note that the entries of the transition matrix, T, are the probabilities of going from one state to another.

The general form of a two-state transition matrix is the following matrix, where p_{ij} is the probability of moving from state j to state i. For example, p_{21} is the probability of moving from state 1 to state 2.

$$T = \begin{array}{c} \text{Current state} \\ \begin{array}{cc} \text{State 1} & \text{State 2} \end{array} \\ \left[\begin{array}{cc} p_{11} & p_{12} \\ & \\ p_{21} & p_{22} \end{array}\right] \end{array} \begin{array}{l} \text{State 1} \\ \\ \text{State 2} \end{array} \quad \text{Next state.}$$

A **stochastic** matrix (p_{ij}) is a square matrix whose entries are probabilities—that is, $0 \le p_{ij} \le 1$, and the sum of components of each column is 1. For example, the following matrices are stochastic:

$$\begin{bmatrix} 0.94 & 0.02 \\ 0.06 & 0.98 \end{bmatrix},$$

$$\begin{bmatrix} 0.8 & 0 \\ 0.2 & 1 \end{bmatrix},$$

$$\begin{bmatrix} 3/4 & 2/5 \\ 1/4 & 3/5 \end{bmatrix},$$

$$\begin{bmatrix} 1 & 0.25 & 0 \\ 0 & 0.5 & 0.5 \\ 0 & 0.25 & 0.5 \end{bmatrix},$$

but the following matrices are not stochastic

$$\begin{bmatrix} 0.82 & 0.6 \\ 0.15 & 0.4 \end{bmatrix},$$

$$\begin{bmatrix} 0 & 1/2 \\ 1 & 3/4 \end{bmatrix},$$

$$\begin{bmatrix} 1 & 0.25 & 0.1 \\ 0 & 0.5 & 0.5 \\ 0 & 0.25 & 0.5 \end{bmatrix},$$

From equation 4.5, the sequence of the distribution vectors,

$$X_1 = TX_0,$$
$$X_2 = TX_1,$$
$$X_3 = TX_2, \ldots,$$
$$X_n = TX_{n-1}, \ldots,$$

is called a **Markov chain**. Informally a **Markov chain** is a sequence of events in which the outcome of current nth event depends only on outcome of the preceding $(n-1)$th event.

Using equation 4.5 we get

$$X_1 = TX_0.$$
$$X_2 = TX_1 = T(TX_0) = T^2 X_0.$$
$$X_3 = TX_2 = T(T^2 X_0) = T^3 X_0.$$

In general,

$$X_n = T^n X_0. \tag{4.7}$$

Let's explore the long-term behavior of this system. We have

$$X_0 = \begin{bmatrix} 0.30 \\ 0.70 \end{bmatrix}.$$

$$X_1 = TX_0 = \begin{bmatrix} 0.2960 \\ 0.7040 \end{bmatrix}.$$

$$X_2 = T^2 X_0 = \begin{bmatrix} 0.2923 \\ 0.7077 \end{bmatrix}.$$

$$X_{10} = T^{10} X_0 = \begin{bmatrix} 0.2717 \\ 0.7283 \end{bmatrix}.$$

It is clear that the city populations, c_n, decrease and the suburbs populations, s_n, increase. Will this trend continue for large values of n? Let's evaluate the following X_n:

$$X_{20} = \begin{bmatrix} 0.2594 \\ 0.7406 \end{bmatrix}.$$

$$X_{50} = \begin{bmatrix} 0.2508 \\ 0.7492 \end{bmatrix}.$$

$$X_{80} = \begin{bmatrix} 0.2501 \\ 0.7499 \end{bmatrix}.$$

$$X_{90} = \begin{bmatrix} 0.2500 \\ 0.7500 \end{bmatrix}.$$

$$X_{100} = \begin{bmatrix} 0.2500 \\ 0.7500 \end{bmatrix}.$$

It can be concluded that for large values of n, X_n reaches $\begin{bmatrix} 0.25 \\ 0.75 \end{bmatrix}$ and stays without change. The distribution (state) vector $\begin{bmatrix} 0.25 \\ 0.75 \end{bmatrix}$ is called the steady state vector or the equilibrium distribution vector.

A **steady-state distribution vector** (or **equilibrium distribution vector**) of a Markov chain described by a first-order difference equation (such as equation 4.5) is a distribution vector, X_e, such that $X_{n+1} = X_n = X_e$—that is X_e satisfies the following equation:

$$X_e = TX_e. \tag{4.8}$$

Note that the vector X_e is the eigenvector corresponding to the eigenvalue $\lambda = 1$. We will use MATLAB to get the eigenvalues of the transition matrix, T, and the corresponding eigenvectors.

```
>> T=[0.94 0.02; 0.06 0.98];
>> [V D] =eig(T)

V=
   -0.7071  -0.3162
    0.7071  -0.9487

D=
   0.9200         0
        0    1.0000
```

Therefore, the vector $\begin{bmatrix} -0.3162 \\ -0.9487 \end{bmatrix}$ is the eigenvector corresponding to eigenvalue $\lambda = 1$. Note that the eigenvector is in the normalized form. Because the sum of the components of the equilibrium vector must be 1, we multiply the vector by the scalar $-1/(0.3162 + 0.9484) = -1/1.2646$. Therefore,

$$X_e = \begin{bmatrix} -0.3162/-1.2646 \\ -0.9487/-1.2646 \end{bmatrix} = \begin{bmatrix} 0.25 \\ 0.75 \end{bmatrix}.$$

This is the same value we obtained by numerical trials (calculating $X_n = T^n X_0$ for large values of n).

Note that the eigenvector $V = \begin{bmatrix} c \\ s \end{bmatrix}$ corresponding to the eigenvalue $\lambda = 1$ may be obtained by finding the reduced echelon form of the homogenous system $(T-I)V = O$. By entering the transition matrix T and **rref(T – eye(2))**, the reduced echelon form is $\begin{bmatrix} 1 & -0.3333 \\ 0 & 0 \end{bmatrix}$, which implies that $c - 0.3333s = 0$. Because $c + s = 1$, the solution of these two equations is $c = 0.25$ and $s_0 = 0.75$. Consequently, $V = X_e = \begin{bmatrix} 0.25 \\ 0.75 \end{bmatrix}$ is the eigenvector corresponding to the eigenvalue 1.

Let's use the results from Section 3.6 to determine the long-term behavior of the above population movement model. We will use the MATLAB function **[V D] = eig(T)** to obtain the eigenvalues and the corresponding eigenvectors of the transition matrix T, which are

$$\lambda_1 = 1 \quad \text{and} \quad V_1 = \begin{bmatrix} -0.3162 \\ -0.9487 \end{bmatrix} = \begin{bmatrix} 1 \\ 3 \end{bmatrix}.$$

$$\lambda_2 = 0.92 \quad \text{and} \quad V_2 = \begin{bmatrix} -0.7071 \\ 0.7071 \end{bmatrix} = \begin{bmatrix} -1 \\ 1 \end{bmatrix}.$$

From Section 3.6, we have

$$X_0 = c_1 V_1 + c_2 V_2$$
$$X_k = c_1 \lambda_1^k V_1 + c_2 \lambda_2^k V_2,$$

where c_1 and c_2 are constants. To determine these two constants, we substitute the values of $X_0 = \begin{bmatrix} 0.3 \\ 0.7 \end{bmatrix}$, $V_1 = \begin{bmatrix} 1 \\ 3 \end{bmatrix}$, and $V_2 = \begin{bmatrix} -1 \\ 1 \end{bmatrix}$ in the equation $X_0 = c_1 V_1 + c_2 V_2$ and solve for c_1 and c_2. We have

$$\begin{bmatrix} 0.3 \\ 0.7 \end{bmatrix} = c_1 \begin{bmatrix} 1 \\ 3 \end{bmatrix} + c_2 \begin{bmatrix} -1 \\ 1 \end{bmatrix}.$$

$$\begin{bmatrix} 0.3 \\ 0.7 \end{bmatrix} = \begin{bmatrix} 1 & -1 \\ 3 & 1 \end{bmatrix} \begin{bmatrix} c_1 \\ c_2 \end{bmatrix}.$$

$$\begin{bmatrix} c_1 \\ c_2 \end{bmatrix} = \begin{bmatrix} 1 & -1 \\ 3 & 1 \end{bmatrix}^{-1} \begin{bmatrix} 0.3 \\ 0.7 \end{bmatrix} = \begin{bmatrix} 0.25 \\ -0.05 \end{bmatrix}.$$

Thus $c_1 = 0.25$ and $c_2 = -0.05$. The general solution is

$$X_k = 0.25(1)^k \begin{bmatrix} 1 \\ 3 \end{bmatrix} - 0.05(0.92)^k \begin{bmatrix} -1 \\ 1 \end{bmatrix},$$

which allows us to conclude that $X_k \to 0.25 \begin{bmatrix} 1 \\ 3 \end{bmatrix} = \begin{bmatrix} 0.25 \\ 0.75 \end{bmatrix}$ as k becomes very large.

Note that this result is the same as the equilibrium vector $X_e = \begin{bmatrix} 0.25 \\ 0.75 \end{bmatrix}$ we obtained.

4.1.3. Regular Markov Chains

Exploration 4.3: A Regular Markov Chain

We will reconsider the population movement model represented by the matrix difference equation 4.5. We found that the distribution vectors, X_n, approach the steady-state (equilibrium) vector, $X_e = \begin{bmatrix} 0.25 \\ 0.75 \end{bmatrix}$, which means that, in the long run, 25% of the total population lives in the city and 75% lives in the suburbs. Recall that the steady-state vector does not depend on the initial distribution vector, X_0. The distribution vector, X_n, is determined from equation 4.7, $X_n = T^n X_0$. Because X_n approaches X_e, where X_0 is fixed, equation 4.7 suggests that T^n approaches a specific matrix. Here are T^n matrices for various values of n,

$$T = \begin{bmatrix} 0.94 & 0.02 \\ 0.06 & 0.98 \end{bmatrix}.$$

$$T^{10} = \begin{bmatrix} 0.5758 & 0.1414 \\ 0.4242 & 0.8586 \end{bmatrix}.$$

$$T^{20} = \begin{bmatrix} 0.3915 & 0.2028 \\ 0.6085 & 0.7972 \end{bmatrix}.$$

$$T^{40} = \begin{bmatrix} 0.2767 & 0.2411 \\ 0.7233 & 0.7589 \end{bmatrix}.$$

$$T^{70} = \begin{bmatrix} 0.2522 & 0.2493 \\ 0.7478 & 0.7507 \end{bmatrix}.$$

$$T^{100} = \begin{bmatrix} 0.2522 & 0.2499 \\ 0.7498 & 0.7501 \end{bmatrix}.$$

$$T^{110} = \begin{bmatrix} 0.2501 & 0.2500 \\ 0.7499 & 0.7500 \end{bmatrix}.$$

$$T^{120} = \begin{bmatrix} 0.2500 & 0.2500 \\ 0.7500 & 0.7500 \end{bmatrix}.$$

$$T^{121} = \begin{bmatrix} 0.2500 & 0.2500 \\ 0.7500 & 0.7500 \end{bmatrix}.$$

It is clear that

$$T^n = \begin{bmatrix} 0.94 & 0.02 \\ 0.06 & 0.98 \end{bmatrix}^n \to \begin{bmatrix} 0.25 & 0.25 \\ 0.75 & 0.75 \end{bmatrix}.$$

Let $X_0 = \begin{bmatrix} c_0 \\ s_0 \end{bmatrix}$ be any initial distribution vector with $c_0 + s_0 = 1$. We have

$$T^n X_0 \to \begin{bmatrix} 0.25 & 0.25 \\ 0.75 & 0.75 \end{bmatrix} \begin{bmatrix} c_0 \\ s_0 \end{bmatrix} = \begin{bmatrix} 0.25c_0 + 0.25s_0 \\ 0.75c_0 + 0.75s_0 \end{bmatrix} = \begin{bmatrix} 0.25(c_0 + s_0) \\ 0.75(c_0 + s_0) \end{bmatrix} = \begin{bmatrix} 0.25 \\ 0.75 \end{bmatrix} = X_e.$$

This allows us to conclude that for any initial distribution vector X_0, the distribution vectors X_n approach (converge to) the steady-state vector X_e.

Do the distribution vectors X_n of a Markov chain always approach a steady-state vector, X_e? The following example shows that this is not the case.

Example 4.1

$$\text{Let } T = \begin{bmatrix} 0 & 1 \\ 1 & 0 \end{bmatrix} \quad \text{and} \quad X_0 = \begin{bmatrix} 0.4 \\ 0.6 \end{bmatrix}.$$

We have

$$T^2 = \begin{bmatrix} 0 & 1 \\ 1 & 0 \end{bmatrix} \begin{bmatrix} 0 & 1 \\ 1 & 0 \end{bmatrix} = \begin{bmatrix} 1 & 0 \\ 0 & 1 \end{bmatrix} = I.$$

$$T^3 = T^2 T = IT = T.$$

In general,

$$T^2 = T^4 = T^6 = \ldots = I.$$

and

$$T = T^3 = T^5 = \dots = \begin{bmatrix} 0 & 1 \\ 1 & 0 \end{bmatrix}.$$

We have

$$X_1 = TX_0 = \begin{bmatrix} 0 & 1 \\ 1 & 0 \end{bmatrix}\begin{bmatrix} 0.4 \\ 0.6 \end{bmatrix} = \begin{bmatrix} 0.6 \\ 0.4 \end{bmatrix}.$$

$$X_2 = T^2 X_0 = IX_0 = X_0 = \begin{bmatrix} 0.4 \\ 0.6 \end{bmatrix}.$$

Therefore,

$$X_0 = X_2 = X_4 = \dots = \begin{bmatrix} 0.4 \\ 0.6 \end{bmatrix}.$$

$$X_1 = X_3 = X_5 = \dots = \begin{bmatrix} 0.6 \\ 0.4 \end{bmatrix}.$$

Obviously, the distribution vectors oscillate indefinitely between the two distribution vectors $\begin{bmatrix} 0.4 \\ 0.6 \end{bmatrix}$ and $\begin{bmatrix} 0.6 \\ 0.4 \end{bmatrix}$. Consequently, the system does not approach any steady-state vector.

Definition
A transition matrix of a Markov chain is called **regular** if for some integer k all the entries of T^k are positive. A Markov chain with a regular transition matrix is called a **regular Markov chain**.

Example 4.2

i. The transition matrix $A = \begin{bmatrix} 0.94 & 0.02 \\ 0.06 & 0.98 \end{bmatrix}$ is regular because all the entries are positive. So the chain $X_{n+1} = AX_n$ is a regular Markov chain.

ii. The stochastic matrix $B = \begin{bmatrix} 0 & 0.4 \\ 1 & 0.6 \end{bmatrix}$ is regular because

$$B^2 = \begin{bmatrix} 0 & 0.4 \\ 1 & 0.6 \end{bmatrix}\begin{bmatrix} 0 & 0.4 \\ 1 & 0.6 \end{bmatrix} = \begin{bmatrix} 0.4 & 0.24 \\ 0.6 & 0.76 \end{bmatrix},$$

where all entries of B^2 are positive.

Regular Markov Chains

Let's summarize the previous results about regular Markov chains If T is a regular transition matrix of a regular Markov chain in the form, $X_{n+1} = TX_n$, then

1. For any initial distribution vector, X_0, for large values of n, the distribution vectors $X_n = T^n X_0$ approach a unique steady-state vector, X_e.
2. The steady-state vector, X_e, can be found by determining the eigenvector of T corresponding to the eigenvalue 1—that is solving the equation $TX_e = X_e$ and taking into consideration the fact that the sum of the elements of X_e is 1.
3. For large values of n, T^n approach a steady-state matrix, L, in which each column is X_e.

iii. Consider the stochastic matrix $C = \begin{bmatrix} 0 & 1 \\ 1 & 0 \end{bmatrix}$. We have

$$C^2 = \begin{bmatrix} 0 & 1 \\ 1 & 0 \end{bmatrix}\begin{bmatrix} 0 & 1 \\ 1 & 0 \end{bmatrix} = \begin{bmatrix} 1 & 0 \\ 0 & 1 \end{bmatrix} = I.$$

$$C^3 = C^2C = IC = C.$$

$$C^4 = C^2C^2 = II = I.$$

In general,

$$C^3 = C^5 = C^7 = \ldots = C$$

and

$$C^2 = C^4 = C^6 = \ldots = I.$$

Note that not all entries of C and I are positive. Because $C^n = C$ if n is odd and $C^n = I$ if n is even, C is not regular.

4.1.4. Genetics Modeling

We investigate the use of matrices in genetics. Consider a population of animals or plants. Each individual in the population possesses two genes. The set of the two genes are A and a. So each individual has one of these combinations: AA, Aa, or aa. Note that Aa is the same as aA. Each individual inherits one gene from each of its parents' pairs of genes with equal likelihood.

The individual's pair of genes is called its **genotype**, which is responsible for the individual's characteristics. For example, in snapdragons genotypes AA, Aa, and aa

produce red, pink, and white flowers, respectively. In humans, the genotypes *AA* and *Aa* produce brown eyes, whereas the genotype *aa* creates blue eyes. Guinea pigs of genotypes *AA* or *Aa* have long hair, whereas those with the *aa* genotype have short hair.

Genetics is important in many areas. Because some diseases are inherited, genetics is important in medicine. Genetic principles are used in plant and animal breeding.

Exploration 4.4: A Genetics Model

A farmer has a population of plants having all three genotypes *AA*, *Aa*, and *aa*. Assume that the farmer breeds all the plants with a plant of genotype *AA*. When both parents have the *AA* genotype, each offspring also is of genotype *AA* with a probability 1. So the probabilities of offspring with *Aa* and *aa* are 0 and 0, respectively.

Suppose the first parent is *AA* and the second parent is *Aa*. Taking one gene from each parent, there are four possibilities for each offspring: *AA*, *Aa*, *AA*, and *Aa* with the probability of $\frac{1}{4}$ for each genotype. Therefore, the probability of each offspring having the *AA*, *Aa*, or *aa* genotype is $\frac{1}{2}$, $\frac{1}{2}$, and 0, respectively. Recall that *Aa* is the same as *aA*. Similarly, if the first parent is *AA* and the second parent is *aa*, the probability of the offspring having the *AA*, *Aa*, or *aa* genotype is 0, 1, and 0, respectively. These results can be put in the following chart:

Genotype of each Offspring	Genotype of the First Parent is *AA* — Genotype of the Second Parent: *AA*	*Aa*	*aa*
AA	1	$\frac{1}{2}$	0
Aa	0	$\frac{1}{2}$	1
aa	0	0	0

This breeding process is a Markov chain having the transition matrix

$$T = \begin{array}{c} \begin{matrix} AA & Aa & aa \end{matrix} \\ \begin{bmatrix} 1 & \frac{1}{2} & 0 \\ 0 & \frac{1}{2} & 1 \\ 0 & 0 & 0 \end{bmatrix} \end{array} \begin{matrix} AA \\ Aa. \\ aa \end{matrix} \tag{4.9}$$

Let a_n, b_n, and c_n be the fraction of the plants of genotype *AA*, *Aa*, and *aa* in the *n*th generation, respectively, for $n=0, 1, 2, \ldots$. Let

$$X_n = \begin{bmatrix} a_n \\ b_n \\ c_n \end{bmatrix} \begin{matrix} AA \\ Aa \\ aa \end{matrix} \tag{4.10}$$

be the population vector in the nth generation. This model can be represented by the following first-order matrix difference equation:

$$\begin{bmatrix} a_n \\ b_n \\ c_n \end{bmatrix} = \begin{bmatrix} 1 & \frac{1}{2} & 0 \\ 0 & \frac{1}{2} & 1 \\ 0 & 0 & 0 \end{bmatrix} \begin{bmatrix} a_{n-1} \\ b_{n-1} \\ c_{n-1} \end{bmatrix}$$

or

$$X_n = TX_{n-1}, \quad n = 1, 2, \ldots. \tag{4.11}$$

Note that T is a stochastic matrix.

Assume that the farmer will start this breeding plan with an initial population consisting of equal fractions of each genotype. That is,

$$X_0 = \begin{bmatrix} \frac{1}{3} \\ \frac{1}{3} \\ \frac{1}{3} \end{bmatrix} \begin{matrix} AA \\ Aa. \\ aa \end{matrix}$$

The solution of equation 4.11 is

$$X_n = T^n X_0.$$

Using MATLAB, we find

$$X_1 = \begin{bmatrix} 0.5 \\ 0.5 \\ 0 \end{bmatrix}.$$

$$X_2 = \begin{bmatrix} 0.75 \\ 0.25 \\ 0 \end{bmatrix}.$$

$$X_3 = \begin{bmatrix} 0.875 \\ 0.125 \\ 0 \end{bmatrix}.$$

$$X_4 = \begin{bmatrix} 0.9375 \\ 0.0625 \\ 0 \end{bmatrix}.$$

$$X_5 = \begin{bmatrix} 0.9688 \\ 0.0313 \\ 0 \end{bmatrix}, \ldots.$$

$$X_{10} = \begin{bmatrix} 0.9990 \\ 0.0010 \\ 0 \end{bmatrix}.$$

$$X_{12} = \begin{bmatrix} 0.9998 \\ 0.0002 \\ 0 \end{bmatrix}.$$

$$X_{13} = \begin{bmatrix} 0.9999 \\ 0.0001 \\ 0 \end{bmatrix}.$$

$$X_{15} = \begin{bmatrix} 1.0000 \\ 0.0000 \\ 0 \end{bmatrix}.$$

$$X_{16} = \begin{bmatrix} 1.0000 \\ 0.0000 \\ 0 \end{bmatrix}.$$

Intuitively, we can conclude that the percent of plants of genotype AA increases and reaches 1, and genotype Aa decreases and reaches 0. Note that the genotype aa disappears from the first generation. This means that all snapdragons will have only red flowers, while the pink and white flowers will disappear.

The last result can be symbolized by

$$X = \lim_{n\to\infty} X_n = \begin{bmatrix} 1 \\ 0 \\ 0 \end{bmatrix}.$$

Exploration 4.5: A Psychology Model

A psychologist conducts an experiment on rats. She puts a rat in a cage with three rooms labeled 1, 2, and 3, as shown in Figure 4.5. The rats are trained to select a door at random and move from one room to another when a bell is rung.

i. Build the transition matrix, T, of the given Markov chain.
ii. If a rat is initially in room 2, what is the probability that this rat will be in room 1 after the bell has rung three times?

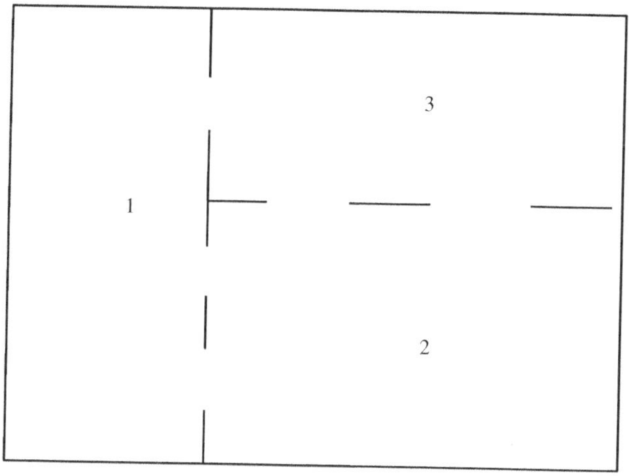

FIGURE 4.5. A cage with three compartments and five doors.

iii. Determine the steady-state vector and interpret it. What is the long-term behavior of this Markov chain?

Discussion

i. To build the transition matrix, T, of this Markov chain, calculate the probabilities of moving from one room to another. Note that p_{ij} is the probability of moving from room j to room i.

$$\begin{array}{c} \text{From} \\ \begin{array}{ccc} R1 & R2 & R3 \end{array} \end{array}$$

$$T = \begin{bmatrix} p_{11} & p_{12} & p_{13} \\ p_{21} & p_{22} & p_{23} \\ p_{31} & p_{32} & p_{33} \end{bmatrix} \begin{array}{l} R1 \\ R2 \;\text{To.} \\ R3 \end{array}$$

For example the first column of T is the probabilities of moving from room 1 (R1) to room 1, $p_{11}=0$, moving from room 1 to room 2 (R2), $p_{21}=\frac{2}{3}$, and moving from room 1 to room 3 (R3), $p_{31}=\frac{1}{3}$. The transition matrix is

$$\begin{array}{c} \text{From} \\ \begin{array}{ccc} R1 & R2 & R3 \end{array} \end{array}$$

$$T = \begin{bmatrix} 0 & \frac{1}{2} & \frac{1}{3} \\ \frac{2}{3} & 0 & \frac{2}{3} \\ \frac{1}{3} & \frac{1}{2} & 0 \end{bmatrix} \begin{array}{l} R1 \\ R2 \;\text{To,} \\ R3 \end{array}$$

and the Markov chain is represented by the matrix difference equation $X_{n+1}=TX_n$.

ii. If the rat is initially in room 2, then the initial state vector is

$$X_0 = \begin{bmatrix} 0 \\ 1 \\ 0 \end{bmatrix}.$$

After the bell has rung three times, we are looking for X_3. Because $X_n = T^n X_0$, we get

$$X_3 = T^3 X_0 = \begin{bmatrix} 0.2222 & 0.3889 & 0.2593 \\ 0.5185 & 0.2222 & 0.5185 \\ 0.2593 & 0.3889 & 0.2222 \end{bmatrix} \begin{bmatrix} 0 \\ 1 \\ 0 \end{bmatrix} = \begin{bmatrix} 0.3889 \\ 0.2222 \\ 0.3889 \end{bmatrix}.$$

Therefore, the probability that the rat is in room 1 after the bell has rung three times is 0.3889.

iii. To find the steady-state vector, X, we need to solve the matrix equation $TX = X$ or the homogeneous system $(T - I)X = O$.

We will use MATLAB command **rref(T – eye(3))**. The output is

$$\begin{bmatrix} 1 & 0 & -1 \\ 0 & 1 & -\frac{4}{3} \\ 0 & 0 & 0 \end{bmatrix}.$$

This implies that $x_1 = t$, $x_2 = \frac{4}{3}t$, and $x_3 = t$. Because the vector X is a probability vector,

$$x_1 + x_2 + x_3 = 1.$$

$$t + \frac{4}{3}t + t = 1.$$

$$\frac{10}{3}t = 1.$$

Therefore, $t = \frac{3}{10}$ and $X = \begin{bmatrix} \frac{3}{10} \\ \frac{4}{10} \\ \frac{3}{10} \end{bmatrix}$. The implementation of this steady-state vector is that the long-term probabilities that the rat is in room 1, room 2, or room 3 are $\frac{3}{10}$, $\frac{4}{10}$, and $\frac{3}{10}$, respectively.

Note that this result can be obtained by using the MATLAB function [V D]=eig(T) to obtain the eigenvalues of the transition matrix, T, and the corresponding eigenvectors. The eigenvalues are $\lambda_1 = 1$, $\lambda_2 = -0.3333$, and $\lambda_3 = -0.6667$ and the corresponding eigenvectors (in normalized form) are $V_1 = \begin{bmatrix} -0.5145 \\ -0.6860 \\ -0.5145 \end{bmatrix}$, $V_2 = \begin{bmatrix} -0.7171 \\ 0 \\ 0.7171 \end{bmatrix}$, and $V_3 = \begin{bmatrix} 0.4082 \\ -0.8165 \\ 0.4082 \end{bmatrix}$, respectively.

The Solution of the system is

$$X_k = c_1\lambda_1^k V_1 + c_2\lambda_2^k V_2 + c_3\lambda_3^k V_3,$$

where c_1, c_2, and c_3 are constants. Because the dominant eigenvalue is $\lambda_1 = 1$,

$$X_k \to c_1 \begin{bmatrix} -0.5145 \\ -0.6860 \\ -0.5145 \end{bmatrix}.$$

Because the sum of components of X_k should equal 1, we multiply X_k by the scalar

$$\frac{1}{c_1(-0.5145 - 0.6860 - 0.5145)} = \frac{-1}{1.7150c_1}.$$

Therefore

$$X_k \to \begin{bmatrix} 0.3 \\ 0.4 \\ 0.3 \end{bmatrix}.$$

4.1.5. Exercises 4.1

In Exercises 1–4, find the distribution vectors X_1, X_2, and X_3 of the Markov chain defined by the transition matrix, T, and the initial distribution vector, X_0.

1. $T = \begin{bmatrix} 0.6 & 0.3 \\ 0.4 & 0.7 \end{bmatrix}$, $X_0 = \begin{bmatrix} 0.4 \\ 0.6 \end{bmatrix}$.

2. $T = \begin{bmatrix} 0.92 & 0.02 \\ 0.08 & 0.98 \end{bmatrix}$, $X_0 = \begin{bmatrix} 0.3 \\ 0.7 \end{bmatrix}$.

3. $T = \begin{bmatrix} \frac{1}{2} & \frac{1}{4} & 0 \\ \frac{1}{2} & \frac{1}{2} & \frac{1}{2} \\ 0 & \frac{1}{4} & \frac{1}{2} \end{bmatrix}$, $X_0 = \begin{bmatrix} 0.50 \\ 0.30 \\ 0.20 \end{bmatrix}$.

4. $T = \begin{bmatrix} 0 & \frac{1}{2} & \frac{1}{2} \\ \frac{1}{2} & 0 & \frac{1}{2} \\ \frac{1}{2} & \frac{1}{2} & 0 \end{bmatrix}, \quad X_0 = \begin{bmatrix} 1 \\ 0 \\ 0 \end{bmatrix}.$

In Exercises 5–10, you are given stochastic matrices. Determine whether the stochastic matrix is regular.

5. $\begin{bmatrix} 0.6 & 1 \\ 0.4 & 0 \end{bmatrix}.$

6. $\begin{bmatrix} 0 & 0.1 \\ 1 & 0.9 \end{bmatrix}.$

7. $\begin{bmatrix} 1 & 0.3 \\ 0 & 0.7 \end{bmatrix}.$

8. $\begin{bmatrix} 0 & 0.2 & 0 \\ 0 & 0.5 & 1 \\ 1 & 0.3 & 0 \end{bmatrix}.$

9. $\begin{bmatrix} 1 & 0 & 0.4 \\ 0 & 1 & 0.3 \\ 0 & 0 & 0.3 \end{bmatrix}.$

10. $\begin{bmatrix} 1 & 0.2 & 0 \\ 0 & 0.6 & 0 \\ 0 & 0.2 & 1 \end{bmatrix}.$

In Exercises 11–14, regular stochastic matrices are given.

A. Find the steady-state distribution vector by solving the matrix equation $TX = X$.

B. Find the eigenvalues and the corresponding eigenvectors of the transition matrix. Then determine the long-term behavior of the Markov chain.

11. $T = \begin{bmatrix} 0.6 & 0.3 \\ 0.4 & 0.7 \end{bmatrix}.$

12. $T = \begin{bmatrix} 0.92 & 0.02 \\ 0.08 & 0.98 \end{bmatrix}.$

13. $T = \begin{bmatrix} \frac{1}{2} & \frac{1}{4} & 0 \\ \frac{1}{2} & \frac{1}{2} & \frac{1}{2} \\ 0 & \frac{1}{4} & \frac{1}{2} \end{bmatrix}.$

14. $T = \begin{bmatrix} 0 & \frac{1}{2} & \frac{1}{2} \\ \frac{1}{2} & 0 & \frac{1}{2} \\ \frac{1}{2} & \frac{1}{2} & 0 \end{bmatrix}, \quad X_0 = \begin{bmatrix} 1 \\ 0 \\ 0 \end{bmatrix}.$

15. Consider a population movement between a certain city and its surrounding suburbs. Assume that

i. The people who move from the city go to the suburbs, and the people who move from the suburbs go to the city,

ii. During a year, the total population in the city and its surrounding suburb is fixed.

iii. The demographic studies showed that 7% of the city population moves to suburb and 2% of the suburb population moves to the city.

A. Model this Markov chain.

B. Use MATLAB to iterate the model with different initial distribution vectors and graph the city's and suburb's populations vs. time. For simplicity use fractions of the populations rather than the actual populations in the distribution vectors. Can you predict the long-term behavior of the system?

C. Determine the steady state vector.

D. Find the eigenvalues of the transition matrix and the corresponding eigenvectors. Use the eigenvalues and the eigenvectors to determine the long-term behavior of the system.

16. In a certain species of roses the plants have red, pink, or white flowers. The color of the flowers is governed by the plant's genotype. The plants of genotype *AA*, *Aa*, and *aa* produce red, pink, and white flowers, respectively. Assume that the farmer breeds all the plants with plants of type *Aa*.

A. Find the transition matrix, *T*, of the Markov chain $X_{n+1} = TX_n$.

B. Assuming that the initial distribution vector is $\begin{bmatrix} 0.25 \\ 0.50 \\ 0.25 \end{bmatrix}$, use MATLAB to find X_n, $n = 1, 2, 3, \ldots, 14$. Do you see a pattern in the distribution vectors? If there is one, describe it.

C. Find a steady-state distribution vector.

D. Find the eigenvalues and corresponding eigenvectors of *T* and use them to determine the long-term behavior of the Markov chain.

17. A psychologist conducts an experiment on a rat in her lab. She puts a mouse in a T-maze, where there are two choices at the T-junction. If the mouse turns right it gets cheese and if it turns left it gets a mild electrical shock. The psychologist records the outcome of each experiment. The observations show that at the first (initial) trial the rat turns right and left with equal probability (50%). If the mouse at a trial turns right, at the next trial the probabilities that the mouse turns right or left are 0.85 and 0.15, respectively. If the mouse turns left at a trial, the probabilities to turn right or left are 0.9 and 0.1, respectively.

A. Use a probability tree to describe this Markov chain.

B. Represent this Markov chain by a matrix equation with a transition matrix, T, and distribution vector, X. Determine the initial vector, X_0.

C. What is the probability that a mouse will turn left on the third trial?

D. What is the long-term behavior of this Markov chain?

18. A psychologist conducts an experiment on rats. She puts a rat in a cage having three rooms labeled 1, 2, and 3, as shown in Figure 4.6. The rats are trained to select a door at random and move from one room to another when a bell is rung.

A. Build the transition matrix, T, of the given Markov chain.

B. If a rat is initially in room 3, what is the probability that this rat will be in room 2 after the bell has rung five times?

C. What is the long-term behavior of this Markov chain? Determine the steady-state vector and interpret it.

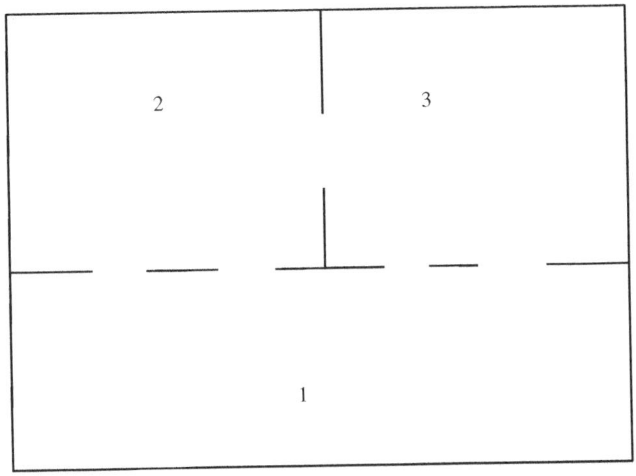

FIGURE 4.6. A cage with three rooms and five doors.

4.2. AGE-STRUCTURED POPULATION MODELS

In this section we investigate the population dynamics of the females of a species that is divided into age classes. This type of model is called a **Leslie's age-structured population model**. Patrick Holt Leslie (1900–1974) was one of the first scientists who used matrix theory to study population dynamics and introduced these models in the 1940s.

Let's explore Leslie's models.

4.2.1. Exploration 4.6

Consider a certain species that lives for at most 3 years. We will group the female population into three age classes: the first age class (0–1 year), the second class (1–2 years), and the third class (2–3 years). Assume that the species has the following characteristics:

i. The survival rate of the species in the first class is 60% (i.e., the probability of an individual in the first class to become a member of the second class is 0.6). The survival rate of the second class is 40%. Because the maximum life span is 3 years, the survival rate of the third class is 0.

ii. The species in the first class do not produce offspring. Each member of the second class produces an average of 6 females. The average number of offspring produced by a member of the third class is $\frac{10}{3}$ (this means that every 3 females produce 10 female offspring).

Model this situation.

Discussion

Let a_n, b_n, and c_n be the population of the female species in the first, second, and third age classes after n years, respectively. This situation can be modeled by the following linear system of difference equations:

$$\begin{aligned} a_{n+1} &= 0a_n + 6b_n + \frac{10}{3}c_n. \\ b_{n+1} &= 0.6a_n. \\ c_{n+1} &= \quad 0.4b_n. \end{aligned} \tag{4.12}$$

This system of difference equations can be represented by the following first-order linear homogenous matrix difference equation:

$$X_{n+1} = LX_n, \tag{4.13}$$

where

$$X_n = \begin{bmatrix} a_n \\ b_n \\ c_n \end{bmatrix}$$

and

$$L = \begin{bmatrix} 0 & 6 & 10/3 \\ 0.6 & 0 & 0 \\ 0 & 0.4 & 0 \end{bmatrix}. \tag{4.14}$$

The vector X_n is the age distribution vector after n years. The matrix L is called **Leslie's matrix**, and this model is called **Leslie's age-structured population model**. We know that the analytical solution of equation 4.13 is

$$X_n = L^n X_0,$$

where $X_0 = \begin{bmatrix} a_0 \\ b_0 \\ c_0 \end{bmatrix}$ is the initial distribution vector.

4.2.2. Exploration 4.7

Consider Leslie's model equations 4.13 and 4.14 with two different initial distribution vectors:

i. $X_0 = \begin{bmatrix} 20 \\ 10 \\ 2 \end{bmatrix}$.

ii. $X_0 = \begin{bmatrix} 50 \\ 15 \\ 3 \end{bmatrix}$.

For each case, find X_n, $n = 0, 1, 2, \ldots, 10$. If there is a pattern, describe it.

Discussion

You may use one of the following MATLAB codes (M-files) to calculate X_n and graph the populations of a_k, b_k, and c_k vs. time in years.

Version 1: A script file called Leslie

% **Leslie** is a script file for Leslie age-structured population dynamic
% The system is represented by X(k) = LX(k-1), where L is Leslie matrix
% Given Leslie matrix and initial distribution vector of a(k), b(k), and c(k)
% This script file graphs a(k), b(k), and c(k) vs. time

```
X= [20; 10; 2];                         % Initial vector X(0)
L= [0 6 10/3;0.6 0 0;0 0.4 0];          % The Leslie matrix, L
P= [X];                                 % P is matrix of 3 rows. X(0) is the
                                        % 1st column of P
T=0:8;                                  % T is the time array 0, 1, 2,..., 8
for k = 1:8                             % Iteration from k = 1 to k=8
  X= L*X;                               % Calculate X(k) = LX(k-1)
  P= [P X];                             % Add the column X(k) to the matrix P
end;
plot(T, P(1, :), 'ko', T, P(2, :), 'k*', T, P(3, :),
'k.');                                  % Plot a(k), b(k), and c(k) vs. time
hold on;
plot(T, P(1, :), 'k', T, P(2, :), 'k', T, P(3, :), 'k');
xlabel('Time n in years');
ylabel('Populations a(n), b(n), and c(n)');
hold off
```

Version 2: A function file called LeslieFunc

```
function [T, P]= LeslieFunc(L, X0, n)
% LeslieFunc is a function to iterate a Leslie system represented by
%     X(n+1) = LX(n)
% The function inputs are
%     L=Lislie matrix
%     X0=initial population vector, where X0=[a(0); b(0); c(0)]; and
%     n=number of time periods
% The function outputs are
%     T=sequence of time
%     P=Matrix of population vectors
%     The graph of a(n), b(n), and c(n) vs. time n

T=0:n;
X=X0;
P= [X0];
for k=1:n
  X=L*X;
  P= [P X];
end;
```

```
plot(T,P(1,:),'ko', T,P(2,:),'k*', T, P(3,:),'k.');
hold on;
plot(T,P(1,:),'k', T,P(2,:),'k', T, P(3,:),'k');
xlabel('Time n in years');
ylabel('populations a(n), b(n), and c(n)');
hold off;
```

To call the function LeslieFunc for $L = \begin{bmatrix} 0 & 6 & 10/3 \\ 0.6 & 0 & 0 \\ 0 & 0.4 & 0 \end{bmatrix}$, $X_0 = \begin{bmatrix} 20 \\ 10 \\ 2 \end{bmatrix}$, and $n=8$ we write the following commands:

```
>> L=[0 6 10/3;0.6 0 0; 0 0.4 0];
>> X0=[20; 10; 2];
>> [T,P]=LeslieFunc(L, X0, 8);
```

We recommend version 2 because it is much more convenient to use the function for different values of L, X_0, and n.

i. The following are X_n, $n=0, 1, 2, \ldots, 8$.

$$X_0 = \begin{bmatrix} 20 \\ 10 \\ 2 \end{bmatrix}.$$

$$X_1 = \begin{bmatrix} 66.7 \\ 12 \\ 4 \end{bmatrix}.$$

$$X_2 = \begin{bmatrix} 85.3 \\ 40 \\ 4.8 \end{bmatrix}.$$

$$X_3 = \begin{bmatrix} 256 \\ 51.2 \\ 16 \end{bmatrix}.$$

$$X_4 = \begin{bmatrix} 360.5 \\ 153.6 \\ 20.5 \end{bmatrix}.$$

$$X_5 = \begin{bmatrix} 989.9 \\ 216.3 \\ 61.4 \end{bmatrix}.$$

$$X_6 = \begin{bmatrix} 1,502.7 \\ 593.9 \\ 86.5 \end{bmatrix}.$$

$$X_7 = \begin{bmatrix} 3,851.9 \\ 901.6 \\ 237.6 \end{bmatrix}.$$

$$X_8 = \begin{bmatrix} 6,201.7 \\ 2,311.2 \\ 360.7 \end{bmatrix}.$$

There is no pattern in the growth of the populations of the three age classes, a_n, b_n, and c_n. The graphs of a_n, b_n, and c_n vs. time (in years) are shown in Figure 4.7. However, there is an interesting pattern if we plot the ratio of each class population to the total population.

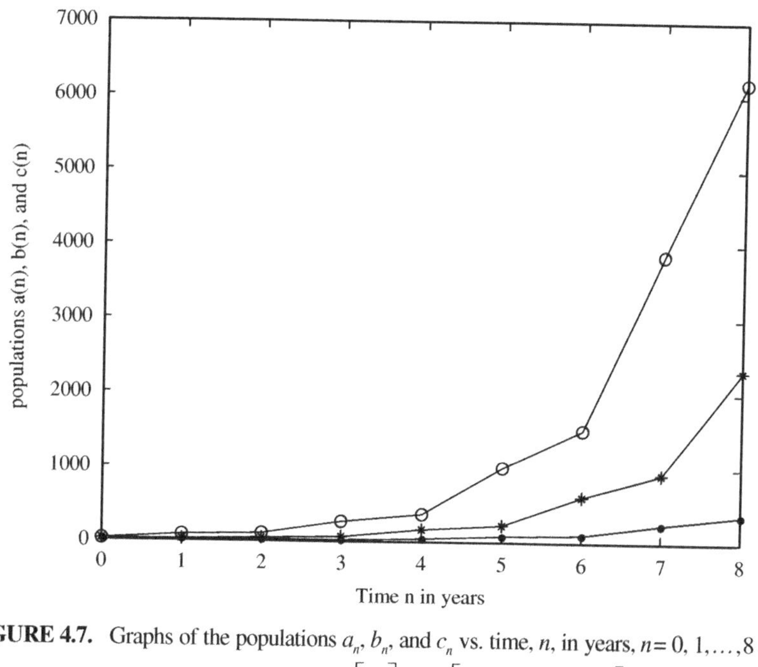

FIGURE 4.7. Graphs of the populations a_n, b_n, and c_n vs. time, n, in years, $n = 0, 1, \ldots, 8$ of the Leslie's model $X_{n+1} = LX_n$, where $X_n = \begin{bmatrix} a_n \\ b_n \\ c_n \end{bmatrix}$, $L = \begin{bmatrix} 0 & 6 & 10/3 \\ 0.6 & 0 & 0 \\ 0 & 0.4 & 0 \end{bmatrix}$, and $X_0 = \begin{bmatrix} 20 \\ 10 \\ 2 \end{bmatrix}$.

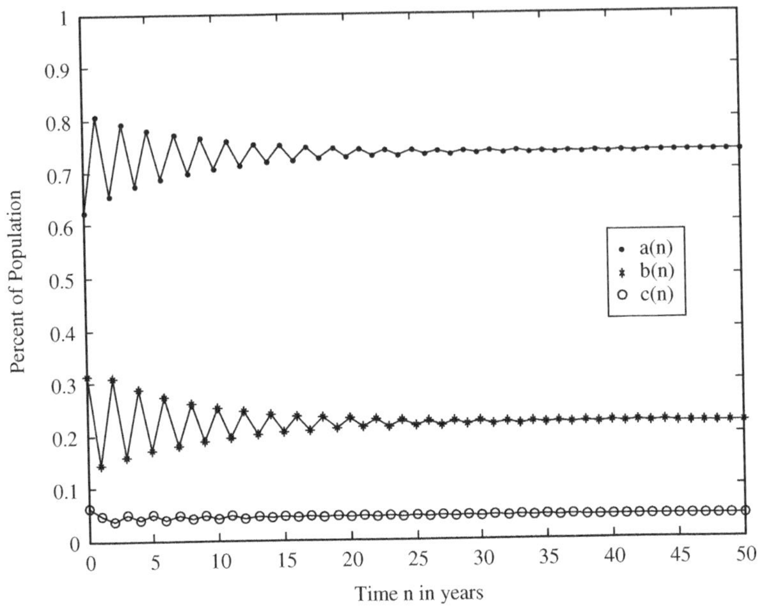

FIGURE 4.8. Graphs of the ratio of each class population a_n, b_n, and c_n to the total population $(a_n+b_n+c_n)$ vs. time, n, in years, $n = 0, 1, \ldots, 50$ of the Leslie's model $X_{n+1}=LX_n$, where $X_n = \begin{bmatrix} a_n \\ b_n \\ c_n \end{bmatrix}$, $L = \begin{bmatrix} 0 & 6 & 10/3 \\ 0.6 & 0 & 0 \\ 0 & 0.4 & 0 \end{bmatrix}$, and $X_0 = \begin{bmatrix} 20 \\ 10 \\ 2 \end{bmatrix}$.

Here is a MATLAB code to graph $\dfrac{a_k}{a_k+b_k+c_k}$, $\dfrac{b_k}{a_k+b_k+c_k}$, and $\dfrac{c_k}{a_k+b_k+c_k}$ vs. time $k = 0, 1, 2, \ldots, 50$. To do that we multiply each distribution vector, X_n, by the factor $\dfrac{1}{a_k+b_k+c_k}$. For example, we need to multiply the vector $X_2 = \begin{bmatrix} 85.3 \\ 40 \\ 4.8 \end{bmatrix}$ by the factor $\dfrac{1}{130.1}$. We get $\begin{bmatrix} 0.6556 \\ 0.3075 \\ 0.0369 \end{bmatrix}$, which means that the first age class is 65.56% of the total population. The populations of the second and third age classes are 30.75% and 3.69% of the total population, respectively. The graphs are shown in Figure 4.8. The percentage of each age class oscillates, and the amplitude of the oscillation decreases and approaches a steady state. The steady states for the percentages of the three age classes are approximately 73.5%, 22%, and 4.4%, respectively.

```
X= [20; 10; 2];
L= [0 6 10/3;0.6 0 0;0 0.4 0];
tp=X(1)+X(2)+X(3);                    % tp=total population
```

```
P=[(1/tp)*X]
T=0:50;
for k =1:50
  X=L*X;
  tp=X(1)+X(2)+X(3)
  P=[P (1/tp)*X];
end;
plot(T, P(1, :), 'k.', T, P(2, :), 'k*', T, P(3, :), 'ko');
hold on;
plot(T, P(1, :), 'k', T, P(2, :), 'k', T, P(3, :), 'k');
xlabel('Time n in years');
ylabel('Percent of Population');
legend('a(n)', 'b(n)', 'c(n)');
hold off;
```

ii. For the second initial distribution vector we have:

$$X_0 = \begin{bmatrix} 50 \\ 15 \\ 3 \end{bmatrix}.$$

$$X_1 = \begin{bmatrix} 100 \\ 30 \\ 6 \end{bmatrix}.$$

$$X_2 = \begin{bmatrix} 200 \\ 60 \\ 12 \end{bmatrix}.$$

$$X_3 = \begin{bmatrix} 400 \\ 120 \\ 24 \end{bmatrix}.$$

$$X_4 = \begin{bmatrix} 800 \\ 240 \\ 48 \end{bmatrix}.$$

$$X_5 = \begin{bmatrix} 1,600 \\ 480 \\ 96 \end{bmatrix}.$$

$$X_6 = \begin{bmatrix} 3,200 \\ 960 \\ 192 \end{bmatrix}.$$

$$X_7 = \begin{bmatrix} 6,400 \\ 1,920 \\ 384 \end{bmatrix}.$$

$$X_8 = \begin{bmatrix} 12,800 \\ 3,840 \\ 768 \end{bmatrix}.$$

The graphs of a_k, b_k, and c_k vs. time are shown in Figure 4.9.

There is a pattern in the dynamics of the distribution vectors $\begin{bmatrix} a_n \\ b_n \\ c_n \end{bmatrix}$. The population of each class increases with the same ratio. We have $a_{n+1}=2a_n$, $b_{n+1}=2b_n$, and $c_{n+1}=2c_n$ or $X_{n+1}=2X_n$. Because $X_{n+1}=LX_n$, we have $LX_n=2X_n$ or $LX=2X$.

This means that $X = \begin{bmatrix} 50 \\ 15 \\ 3 \end{bmatrix}$ (or a scalar multiple of X) is an eigenvector of L corresponding to the eigenvalue $\lambda=2$. This pattern represents a steady-state growth rate.

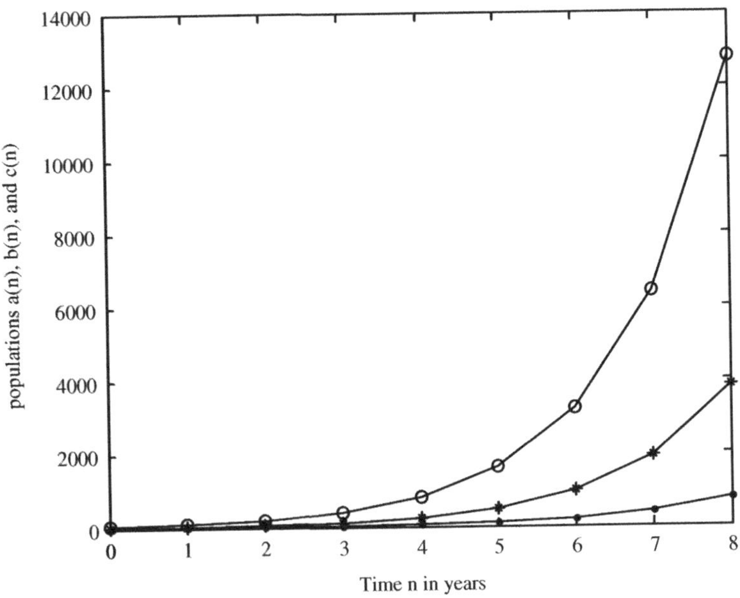

FIGURE 4.9. Graphs of the populations a_n, b_n, and c_n vs. time, n, in years, $n=0, 1, \ldots, 8$ of the Leslie's model $X_{n+1}=LX_n$, where $X_n = \begin{bmatrix} a_n \\ b_n \\ c_n \end{bmatrix}$, $L = \begin{bmatrix} 0 & 6 & 10/3 \\ 0.6 & 0 & 0 \\ 0 & 0.4 & 0 \end{bmatrix}$, and $X_0 = \begin{bmatrix} 50 \\ 15 \\ 3 \end{bmatrix}$.

Another interpretation of the pattern in the vectors, X_n, is that the ratios of a_n, b_n, and c_n to the total population $(a_n+b_n+c_n)$ for all n are 0.7353, 0.2206, and 0.0441, respectively. For example,

$$\frac{1}{50+15+3}\begin{bmatrix}50\\15\\3\end{bmatrix}=\frac{1}{68}\begin{bmatrix}50\\15\\3\end{bmatrix}=\begin{bmatrix}0.7353\\0.2206\\0.0441\end{bmatrix}$$

and

$$\frac{1}{200+60+12}\begin{bmatrix}200\\60\\12\end{bmatrix}=\frac{1}{272}\begin{bmatrix}200\\60\\12\end{bmatrix}=\begin{bmatrix}0.7353\\0.2206\\0.0441\end{bmatrix}.$$

In other words, if $p_n=a_n+b_n+c_n$ is the total population, then

$$a_n = 73.53\% \approx 74\% \quad \text{of} \quad p_n.$$
$$b_n = 22.06\% \approx 22\% \quad \text{of} \quad p_n.$$
$$c_n = 4.41\% \approx 4\% \quad \text{of} \quad p_n.$$

Again this is a steady-state growth rate. The graphs of the percentages of a_n, b_n, and c_n with respect to the total population $(a_n+b_n+c_n)$ are shown in Figure 4.10.

4.2.3. Exploration 4.8

Consider the Leslie age-structured population model 4.12, where $X_{n+1}=LX_n$ and

$$L=\begin{bmatrix}0 & 6 & 10/3\\0.6 & 0 & 0\\0 & 0.4 & 0\end{bmatrix}.$$

i. Find a steady-state growth rate distribution vector if it exists.
ii. Determine the long-term behavior of the system.

Discussion

i. We start by computing the eigenvalues of L. The methods of Section 3.5 may be used to calculate the eigenvalues of L, but we will use the MATLAB function **eig(L)**. The eigenvalues are $\lambda_1=2$, $\lambda_2=-1.7746$, and $\lambda_3=-0.2254$. Only positive eigenvalues have biological meaning in this situation. Consequently, we consider the positive eigenvalue $\lambda_1=2$ and compute the corresponding eigenvector. Let $V=\begin{bmatrix}x_1\\x_2\\x_3\end{bmatrix}$ be the eigenvector of L corresponding to the eigenvalue $\lambda=2$. The MATLAB routine rref(L – 2*eye(3)) returns

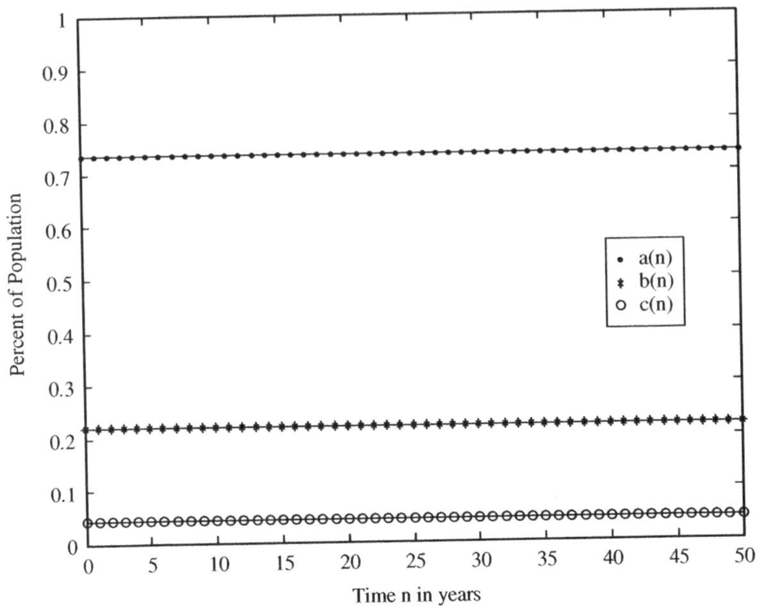

FIGURE 4.10. Graphs of the ratio of each class population a_n, b_n, and c_n to the total population $(a_n+b_n+c_n)$ vs. time, n, in years, $n=0, 1, \ldots, 50$ of the Leslie's model $X_{n+1}=LX_n$, where $X_n=\begin{bmatrix} a_n \\ b_n \\ c_n \end{bmatrix}$, $L=\begin{bmatrix} 0 & 6 & 10/3 \\ 0.6 & 0 & 0 \\ 0 & 0.4 & 0 \end{bmatrix}$, and $X_0=\begin{bmatrix} 50 \\ 15 \\ 3 \end{bmatrix}$.

$$\begin{bmatrix} 1 & 0 & -16.6667 \\ 0 & 1 & -5 \\ 0 & 0 & 0 \end{bmatrix},$$

which allows us to conclude that $x_1=16.6667x_3$ and $x_2=5x_3$. Letting $x_3=t$ (a parameter), the eigenvector is in the parametric form

$$V=t\begin{bmatrix} 16.6667 \\ 5 \\ 1 \end{bmatrix}.$$

To get integer components of V, let $t=3$. Consequently, the eigenvector of L corresponding to $\lambda=2$ is $V=\begin{bmatrix} 50 \\ 15 \\ 3 \end{bmatrix}$. We concluded in Exploration 4.7 that the vector, V, is a steady growth rate distribution vector of Leslie matrix, L.

ii. Because the matrix L has three different real eigenvalues $\lambda_1=2$, $\lambda_2=-1.7746$, and $\lambda_3=-0.2254$, the solution of equation 4.8 is

$$X_k = c_1(\lambda_1)^k V_1 + c_2(\lambda_2)^k V_2 + c_3(\lambda_3)^k V_2,$$

where c_1, c_2, and c_3 are constants and V_1, V_2, V_3 are the corresponding eigenvectors.

Because $\lambda_1 = 2$ is the dominant eigenvalue and $V_1 = \begin{bmatrix} 50 \\ 15 \\ 3 \end{bmatrix}$ is the dominant eigenvector, we have

$$X_k \to c_1 2^k \begin{bmatrix} 50 \\ 15 \\ 3 \end{bmatrix}.$$

Thus X_k increases without bound for large values of k.

Note that in Exploration 4.8 there was only one positive eigenvalue and the corresponding eigenvector had positive components. What if matrix L has more than one positive eigenvalue or no positive eigenvalue? The next theorem, which we state without proof, addresses all these concerns.

Theorem 1

Every Leslie matrix L has a unique positive eigenvalue, λ_1. All the components of the corresponding eigenvector, V_1, are positive. Moreover, if L has any other real or complex eigenvalue, λ_k, then $|\lambda_k| \le \lambda_1$.

4.2.4. Exercises 4.2

In Exercises 1–6 an age-structured population model is in the form $X_{n+1} = LX_n$, where Leslie matrix L and the initial distribution vector X_0 are given. Find X_1, X_2, and X_3.

1. $L = \begin{bmatrix} 0 & 5 \\ 0.8 & 0 \end{bmatrix}, \quad X_0 = \begin{bmatrix} 10 \\ 5 \end{bmatrix}.$

2. $L = \begin{bmatrix} 0 & 3 \\ 0.75 & 0 \end{bmatrix}, \quad X_0 = \begin{bmatrix} 8 \\ 8 \end{bmatrix}.$

3. $L = \begin{bmatrix} 0 & 2 \\ \frac{1}{8} & 0 \end{bmatrix}, \quad X_0 = \begin{bmatrix} 10 \\ 4 \end{bmatrix}.$

4. $L = \begin{bmatrix} 0 & 3 & 8 \\ 1 & 0 & 0 \\ 0 & \frac{1}{4} & 0 \end{bmatrix}, \quad X_0 = \begin{bmatrix} 8 \\ 8 \\ 8 \end{bmatrix}.$

5. $L = \begin{bmatrix} 0 & 3 & 2 \\ 0.5 & 0 & 0 \\ 0 & 0.2 & 0 \end{bmatrix}, \quad X_0 = \begin{bmatrix} 10 \\ 10 \\ 10 \end{bmatrix}.$

6. $L = \begin{bmatrix} 0 & 3 & 2 \\ 0.2 & 0 & 0 \\ 0 & 0.1 & 0 \end{bmatrix}, \quad X_0 = \begin{bmatrix} 20 \\ 20 \\ 20 \end{bmatrix}.$

In Exercises 7–12 do the following.

A. Find the unique positive eigenvalue, λ_1, of the Leslie matrix, L and the corresponding eigenvector, V_1.

B. Find a stable age distribution vector and determine the long-term behavior of Leslie model.

7. Leslie model in Exercise 1.

8. Leslie model in Exercise 2.

9. Leslie model in Exercise 3.

10. Leslie model in Exercise 4.

11. Leslie model in Exercise 5.

12. Leslie model in Exercise 6.

13. Consider the female population of a certain species with three age classes of 2 years' duration. The population has the following characteristics:

i. The survival rate from class 1 to class 2 is 50% and from class 2 to class 3 is 40%. The maximum life span is 6 years.

ii. On average each female in class 2 and class 3 produces four and two female offspring, respectively, while females in class 1 produce no female offspring.

A. Model this situation.

B. Find a stable age distribution vector.

C. For an arbitrary initial distribution vector, $X_0 = \begin{bmatrix} 10 \\ 10 \\ 10 \end{bmatrix}$, calculate the ratio of the population of each class to the total population of the distribution vector for the vectors $X_n = \begin{bmatrix} a_n \\ b_n \\ c_n \end{bmatrix}$, $n = 1, 2, \ldots, 20$. Plot these ratios. What can be concluded from these graphs?

4.3. MODELING WITH SECOND-ORDER LINEAR DIFFERENCE EQUATIONS

In this section we are interested in modeling real-life situations with second-order linear difference equations. We will explore methods to explore such models.

4.3.1. Introduction to Second-Order Difference Equations

In the population models we discussed earlier we assumed that every member of the species in the nth generation contributes to the population in the $(n+1)$st generation. For some species, there is a maturation period, such as baleen whales, for which the nth and $(n+1)$st generations contribute to the $(n+2)$nd generation. Assume for a certain species, two consecutive generations P_n and P_{n+1} simultaneously contribute to the population P_{n+2}. The population can be modeled by the following difference equation:

$$P_{n+2} = \alpha P_{n+1} + \beta P_n,$$

where $\alpha, \beta \geq 0$.

This equation is a second-order linear difference equation because the difference between the highest and the lowest subscripts $(n+2)-n$ is 2. For a first-order difference equation we needed an initial condition to iterate and obtain a numerical solution. Similarly, for a second-order difference equation we need two initial conditions to be able to iterate the difference equation.

There are methods for obtaining a closed-form solution (analytical solutions) of most second-order linear difference equations with constant coefficients in the form

$$ax_{n+2} + bx_{n+1} + cx_n = d_n,$$

where the coefficients a, b, and c are constants and d_n is not a constant because it may depend on n. In this section we will use the following approach in studying models with second-order linear difference equations with constant coefficients.

Convert a second-order difference equation into a system of two first-order difference equations, then represent the system by a first-order linear matrix equation. Find the eigenvalues and the corresponding eigenvector. Utilizing the techniques developed in Section 3.6, the analytical solution may be formed and the system's long-term behavior can be determined. MATLAB may be used to find the eigenvalues and the corresponding eigenvectors as well as iterating and graphing a numerical solution.

We will introduce a technique to convert a second-order difference equation to a system of two first-order difference equations and a third-order difference equation can be converted to three first-order difference equations. In general, any kth-order difference equation can be converted to k first-order difference equations.

Example 4.3

Consider the following linear second-order difference equation

$$x_{n+2} = 2x_{n+1} - x_n,$$

with the initial conditions $x_1 = 1$ and $x_2 = 5$.

i. Find x_3, x_4 and x_5.
ii. Convert the second-order equation into a system of first-order difference equations.
iii. Represent the system by a matrix equation
iv. Use MATLAB to iterate the system and graph x_n vs. n.

Solution

i. Letting $n = 1$, we get

$$x_3 = 2x_2 - x_1 = 2(5) - 1 = 9.$$

Similarly,

$$x_4 = 2x_3 - x_2 = 2(9) - 5 = 13$$
$$x_5 = 2x_4 - x_3 = 2(13) - 9 = 17.$$

Note that the iteration of the same difference equation $x_{n+2} = 2x_{n+1} - x_n$ with different initial conditions produces different sequence of x_n. For instance, letting $x_1 = 2$ and $x_2 = -4$, the terms x_3, x_4, and x_5 as follows:

$$x_3 = 2x_2 - x_1 = 2(-4) - 2 = -10$$
$$x_4 = 2x_3 - x_2 = 2(-10) - 4(-4) = -16.$$
$$x_5 = -22$$

ii. To convert the second-order difference equation $x_{n+2} = 2x_{n+1} - x_n$, we introduce a new dependent variable, y_n,

$$y_n = x_{n+1}.$$

Note that $y_{n+1} = x_{n+2}$. Substitute in the equation for x_{n+1} and x_{n+2} to get

$$y_{n+1} = 2y_n - x_n.$$

Therefore, we have the following system of two first-order linear difference equations:

$$x_{n+1} = y_n.$$
$$y_{n+1} = 2y_n - x_n.$$

We need to convert the initial condition x_1 and x_2 into initial conditions x_1 and y_1. We have $x_1 = 1$. Letting $n=1$ in the equation $x_{n+1} = y_n$, we get $x_2 = y_1$, and because $x_2 = 5$, we obtain $y_1 = 5$. Therefore, the new initial conditions are $x_1 = 1$ and $y_1 = 5$.

Iterating the system with the initial conditions $x_1 = 1$ and $y_1 = 5$, we get

$$x_2 = y_1 = 5$$
$$y_2 = 2y_1 - x_1 = 2(5) - 1 = 9.$$

$$x_3 = y_2 = 9$$
$$y_3 = -x_2 + 2y_2 = -5 + 2(9) = 13.$$

$$x_4 = y_3 = 13$$
$$y_4 = -x_3 + 2y_3 = -9 + 2(13) = 17.$$

Note that x_1, x_2, x_3, and x_4 have the same values obtained by iterating the second-order difference equation.

iii. The system

$$x_{n+1} = y_n$$
$$y_{n+1} = 2y_n - x_n$$

can be written as a first-order matrix difference equation:

$$V_{n+1} = AV_n,$$

where $V_n = \begin{bmatrix} x_n \\ y_n \end{bmatrix}$, $A = \begin{bmatrix} 0 & 1 \\ -1 & 2 \end{bmatrix}$ and $V_0 = \begin{bmatrix} 1 \\ 5 \end{bmatrix}$. The solution of this difference equation is given by

$$V_n = A^n V_0.$$

For example, $V_3 = \begin{bmatrix} 0 & 1 \\ -1 & 2 \end{bmatrix}^3 \begin{bmatrix} 1 \\ 5 \end{bmatrix} = \begin{bmatrix} 13 \\ 17 \end{bmatrix}$, $V_{10} = \begin{bmatrix} 41 \\ 45 \end{bmatrix}$, and $V_{20} = \begin{bmatrix} 81 \\ 85 \end{bmatrix}$.

iv. Here is a possible MATLAB code of a function called **SecondOrderDE** that iterates a second-order difference equation in the form

$$x_{n+2} = ax_{n+1} + bx_n,$$

with the initial conditions x_1 and x_2.

```
function [T, X] = SecondOrderLDE(a, b, x1, x2, k)
% This function iterates the second-order linear difference equation
%   x(n+2) = ax(n + 1) + bx(n) or equivalently x(n) = ax(n-1)+ bx(n-2)
% with the initial values x1 and x2
% The function input (parameters) are:
%   the constants a and b
%   the initial conditions x1 and x2
%   the number of iterations k
% Function output:
%   T=sequence of time
%   X=sequence of x(n)

T=1:k;
X=zeros(1,k);
X(1)=x1; X(2)=x2;
for j=3:k
  X(j)=a*X(j-1)+b*X(j-2);
end;
plot(T, X, 'k.', T, X, 'k');
xlabel('Time periods n');
ylabel('x(n) values');
return;
```

To call the function **SecondOrderDE** for the difference equation $x_{n+2}= 2x_{n+1}-x_n$ with the initial conditions $x_1=1$ and $x_2=5$ for the first 12 values, including the initial values, write the following command:

```
>> [T, X] = SecondOrderLDE(2, -1, 1, 5, 12);
```

The ordered pairs (n, x_n), $n=1, 2, \ldots, 12$ are in the following chart and are graphed in Figure 4.11A.

n	1	2	3	4	5	6	7	8	9	10	11	12
x_n	1	5	9	13	17	21	25	29	33	37	41	45

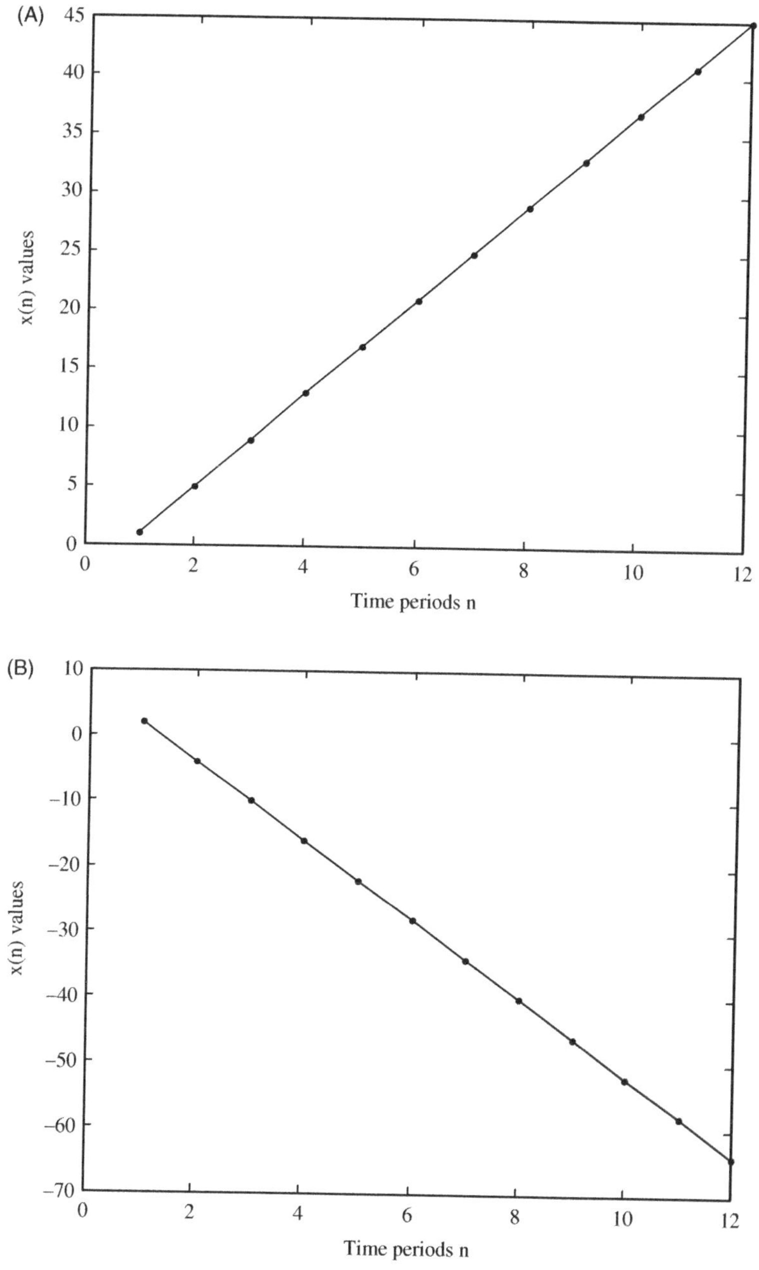

FIGURE 4.11. Graphs of the ordered pairs (n, x_n), $n = 1, 2, \ldots, 12$ of the second-order difference equation $x_{n+2} = 2x_{n+1} - x_n$. **A**. Initial conditions $x_1 = 1$ and $x_2 = 5$. **B**. Initial conditions $x_1 = 2$ and $x_2 = -4$.

To call the function **SecondOrderDE** for the difference equation $x_{n+2}=2x_{n+1}-x_n$ with the initial conditions $x_1=2$ and $x_2=-4$ for the first 12 values, write the following command:

```
>> [T, X] = SecondOrderLDE(2, -1, 2, -4, 12);
```

The ordered pairs (n, x_n), $n=1, 2, \ldots, 12$ are given in the following chart and shown in Figure 4.11B.

n	1	2	3	4	5	6	7	8	9	10	11	12
x_n	2	−4	−10	−16	−22	−28	−34	−40	−46	−52	−58	−64

Example 4.4

Convert the third-order difference equation $x_{n+3}= 2x_{n+2} + 4x_{n+1} -3x_n$, with the initial conditions $x_0=1$, $x_1=3$, and $x_2=-2$, into a system of first-order difference equations.

Solution

Let $y_n=x_{n+2}$ and $z_n=x_{n+1}$. We have $y_{n+1}=x_{n+3}$ and $z_{n+1}=x_{n+2}=y_n$. Consequently, the third-order difference equation is converted to a system of the following three first-order difference equations:

$$\begin{aligned} x_{n+1} &= z_n \\ y_{n+1} &= -3x_n + 2y_n + 4z_n . \\ z_{n+1} &= y_n \end{aligned}$$

The new initial conditions for this system are $x_0=1$, $y_0=-2$, and $z_0=3$.

4.3.2. Model 4.1: Seals Population Dynamics

Consider a population of female seals in a colony with the following traits:

i. The survival rate of female seals is a per year, where $0 \le a \le 1$. Note that $a=1-d$, where d is the seals death rate and $0 \le d \le 1$.

ii. Female seals become sexually mature and therefore become productive on their second year. That is each female seal gives birth to new seals in her second year and every year after.

iii. The fertility rate of a productive female seal is b—that is, every productive female seal gives birth to b female seals per year.

Let's model the population dynamic of female seals and predict its long-term behavior.

Discussion

Let p_n be the population of female seals at year n. Based on the given traits of female seals, the population dynamics of female seals can be modeled by the following difference equation;

$$p_{n+2} = ap_{n+1} + bp_n$$

or

$$p_n = ap_{n-1} + bp_{n-2}, \quad n = 3,4,\ldots,k. \tag{4.15}$$

The initial conditions are p_1 and p_2. Because a and b are constants, equation 4.15 is a second-order linear difference equation with constant coefficients.

The following is a MATLAB function, called **SealsDynamics_2LDE**, which iterates the second-order linear difference equation 4.15 with the initial conditions p_1 and p_2 as well as graphs p_n vs. n. The function accepts a, b, p_1, p_2, and k, the number of iterations. It returns N, a sequence of n; P, a sequence of populations, p_n; and a graph.

```
function [N, P] =SealsDynamics_2LDE(a, b, p1, p2, k)
% This function iterates the second-order difference equation
% p(n) = ap(n-1) + bp(n-2)
% with the initial values p1 and p2
% The function parameters are:
%   the constants a and b
%   the initial conditions p1 and p2
%   the number of ordered pairs k
% Function output:
% N=sequence of time n
% P=sequence of the Seals populations p(n)

N=1:k;
P=zeros (1,k);
P(1) =p1; P(2) =p2;
for j=3:k
  P(j) =a*P(j-1) +b*P(j-2);
end;
plot(N, P, 'k.', N, P, 'k');
xlabel('Time periods n in years');
ylabel('The Seals population p(n)');
return;
```

Let's consider equation 4.15 with $a=0.7$ and $b=0.6$ and with the initial conditions $p_1=50$ and $p_2=52$. To graph (n, P_n) for $n=1, 2, \ldots, 15$, iterate this equation 15 times:

```
>>[N, P] =SealsDynamics_2LDE(0.7, 0.6, 50, 52, 15)
```

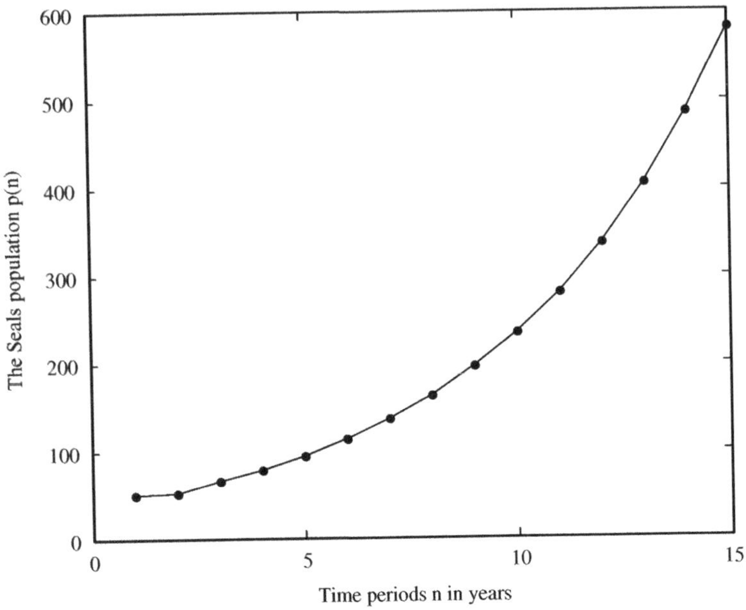

FIGURE 4.12. Graphs of seal population, p_n, vs. time, n, in years, $n = 1, 2, \ldots, 15$ modeled by the second-order difference equation $p_{n+2}=0.7p_{n+1}+0.6p_n$, with the initial conditions $p_1=50$ and $p_2=52$.

The approximated values of the ordered pairs (n, p_n), $n=1, 2, \ldots, 15$ are shown in the following chart:

n	1	2	3	4	5	6	7	8	9	10	11	12	13	14	15
p_n	50	52	66	78	94	113	135	162	195	234	280	337	404	485	582

The graph of p_n vs. n is shown in Figure 4.12.

Another way to investigate this problem is to convert the following second-order difference equation

$$p_{n+2} = 0.7p_{n+1} + 0.6p_n, \qquad p_1 = 50 \quad \text{and} \quad p_2 = 52 \tag{4.16}$$

into a system of two first-order difference equations. Letting $q_n = p_{n+1}$, we obtain the system

$$\begin{aligned} p_{n+1} &= q_n \\ q_{n+1} &= 0.6p_n + 0.7q_n, \end{aligned}$$

with the initial conditions $p_1 = 50$ and $q_1 = 52$.

Note that this system of two first-order linear difference equations can be represented by the following matrix equation:

$$X_{n+1} = TX_n,$$

where

$$X_n = \begin{bmatrix} p_n \\ q_n \end{bmatrix},$$

$$T = \begin{bmatrix} 0 & 1 \\ 0.6 & 0.7 \end{bmatrix},$$

and

$$X_0 = \begin{bmatrix} 50 \\ 52 \end{bmatrix}. \tag{4.17}$$

The eigenvalues and eigenvectors method studied in Chapter 3 may be used to study the long-term behavior of the system.

To determine the eigenvalues and the corresponding eigenvectors, we will use the MATLAB function [V D]=eig(T). The matrix T has two eigenvalues, $\lambda_1 = 1.2$ and $\lambda_2 = -0.5$, and the corresponding eigenvectors are $V_1 = \begin{bmatrix} 0.8333 \\ 1 \end{bmatrix}$ and $V_2 = \begin{bmatrix} -2 \\ 1 \end{bmatrix}$, respectively. From Section 3.6, the solution of the equation is given by

$$X_k = c_1\lambda_1^k V_1 + c_2\lambda_2^k V_2,$$

which simplifies to

$$X_k = c_1(1.2)^k \begin{bmatrix} 0.8333 \\ 1 \end{bmatrix} + c_2(-0.5)^k \begin{bmatrix} -2 \\ 1 \end{bmatrix}, \tag{4.18}$$

where c_1 and c_2 are constants. The values of c_1 and c_2 can be determined by substituting the initial vector, $X_1 = \begin{bmatrix} 50 \\ 52 \end{bmatrix}$, in the equation, with $k = 1$ and solve for c_1 and c_2. We have $c_1 = 45.2947$ and $c_2 = 4.7072$. Consequently, the analytical solution of the system with the given initial conditions is

$$X_k = 45.2947(1.2)^k \begin{bmatrix} 0.8333 \\ 1 \end{bmatrix} + 4.7072(-0.5)^k \begin{bmatrix} -2 \\ 1 \end{bmatrix}.$$

Because $|\lambda_1|=1.2>1$ and $|\lambda_2|=0.5<1$, $X_k \rightarrow \infty$. This result is in agreement with the Figure 4.12, which showed that the p_n values increase exponentially.

4.3.3. Model 4.2: Seals Population Dynamics, Revisited

Consider equation 4.15, with $a=0.2$ and $b=0.8$ and with the initial conditions $p_1=50$ and $p_2=54$. Explore the dynamics of the seal population.

Discussion

The model is represented by the second-order linear difference equation

$$p_{n+2} = 0.2 p_{n+1} + 0.8 p_n, \qquad p_1 = 50 \quad \text{and} \quad p_2 = 54 \tag{4.19}$$

We used the function **SealsDynamics_2LDE** to graph P_n vs. n for the first 40 n by writing the command

```
>> [N, P] = SealsDynamics_2LDE(0.2, 0.8, 50, 54, 40);
```

The graph is shown in Figure 4.13. Before we describe this interesting graph, let's convert the second-order difference equation into a system of two first-order

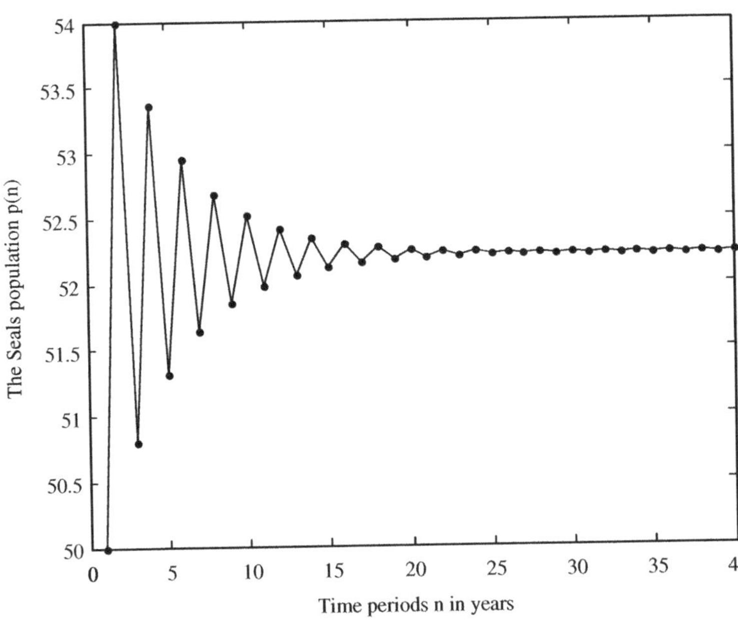

FIGURE 4.13 Graphs of seal population, p_n, vs. time, n, in years, $n=1, 2, \ldots, 40$ modeled by the second-order difference equation $p_{n+2}=0.2p_{n+1}+0.8p_n$, with the initial conditions $p_1=50$ and $p_2=54$.

difference equations and determine the eigenvalues and the corresponding eigenvectors to determine the long-term behavior of the system.

This equation is transformed into the following system of two first-order difference equations

$$p_{n+1} = q_n.$$
$$q_{n+1} = 0.8p_n + 0.2q_n.$$

The matrix representation of this model is

$$X_{n+1} = TX_n,$$

where

$$X_n = \begin{bmatrix} p_n \\ q_n \end{bmatrix}.$$

$$T = \begin{bmatrix} 0 & 1 \\ 0.8 & 0.2 \end{bmatrix}.$$

$$X_1 = \begin{bmatrix} 50 \\ 54 \end{bmatrix}. \tag{4.20}$$

The matrix T has two eigenvalues, $\lambda_1 = 1$ and $\lambda_2 = -0.8$, and the corresponding eigenvectors are $V_1 = \begin{bmatrix} 1 \\ 1 \end{bmatrix}$ and $V_2 = \begin{bmatrix} -1.25 \\ 1 \end{bmatrix}$, respectively. From Section 3.6, the solution of the equation is given by

$$X_k = c_1\lambda_1^k V_1 + c_2\lambda_2^k V_2,$$

which simplifies to

$$X_k = c_1(1)^k \begin{bmatrix} 1 \\ 1 \end{bmatrix} + c_2(-0.8)^k \begin{bmatrix} -1.25 \\ 1 \end{bmatrix}, \tag{4.21}$$

where c_1 and c_2 are constants. Because the dominant eigenvalue is $\lambda_1 = 1$ and the other eigenvalue, $\lambda_2 = -0.8$, is negative with $|\lambda_2| < 1$, the second term of X_k approaches 0 and $X_k \to c_1 \begin{bmatrix} 1 \\ 1 \end{bmatrix}$.

Because $(-0.8)^k$ is positive for even k and is negative for odd k, there is an oscillation in the values of X_k. The amplitude of oscillation is decreasing, and the system reaches the equilibrium value 52.2222.

4.3.4. Model 4.3: Plant Population Dynamics

Consider a plant with a regular life cycle: seeds germinate and grow into seedlings; the seedlings grow and develop roots, stems, and leaves; the plants grow, mature, and produce flowers; the flowers produce seeds; and the plants die. With the seeds, the plant cycle starts again. (Note that this model is based on Reference 5.)

Assume the plant has the following traits (characteristics):

i. Each plant produces a seeds
ii. The probability that a seed survives the winter is b.
iii. The probability that a 1-year-old seed germinates is c.
iv. The probability that a 2-year-old seed germinates is d.
v. The life of the seeds is 2 years, that is, seeds older than 2 years do not germinate.

Discussion

Let's build a model of the plant's population dynamics:

Let p_n be the number of plants in year n. We have

$$p_{n+2} = \text{Plants from 1-year-old seeds} + \text{Plants from 2-year-old seeds}.$$

We will use the given plant's traits to calculate the number of plants from 1-year-old and 2-year-old seeds.

To calculate the plants from 1-year-old seeds we have

The number of plants last year $(n+1) = p_{n+1}$.

The number of seeds produced last year $= ap_{n+1}$.

The number of seeds produced last year and survived the first winter $= bap_{n+1}$.

The number of seeds that germinate among these seeds $= cbap_{n+1}$.

Thus the plants from 1-year-old seeds $= cbap_{n+1}$.

To calculate the plants from 2-year-old seeds we have

The number of plants 2 years ago is p_n.

The number of seeds produced 2 years ago $= ap_n$.

The number of the seeds which survived the first winter $= bap_n$.

The number of these seeds that survived the first winter and did not germinate first year $= (1-c)bap_n$.

The number of these seeds that survived the second winter $= b(1-c)bap_n$.

The number of these seeds that germinate during the second year $= db(1-c)bap_n$.

Thus the plants from 2-year-old seeds $= db(1-c)bap_n$.

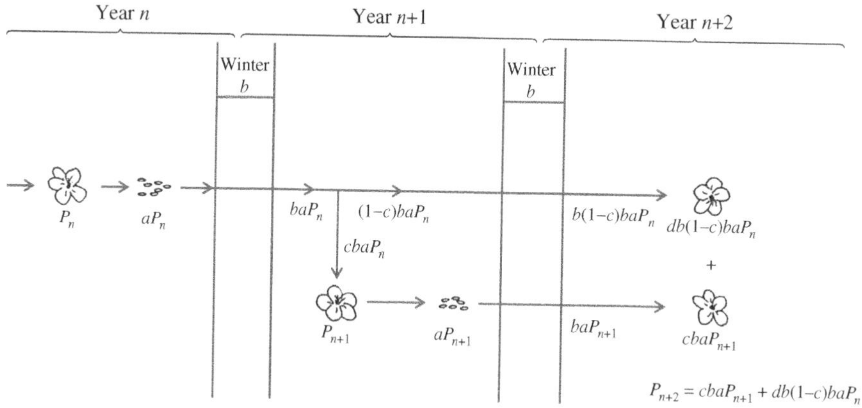

FIGURE 4.14. Propagation of annual plants with the following characteristics: each plant produces a seeds; the probability that a seed survives the winter is b; the probability that a 1-year-old seed germinates is c; the probability that a 2-year-old seed germinates is d; and the life of the seeds is 2 years.

Therefore,

$$p_{n+2} = cbap_{n+1} + db(1-c)bap_n. \tag{4.22}$$

Because a, b, c, and d are constants, the equation is a second-order linear difference equation with constant coefficients.

Letting $\alpha = cba$ and $\beta = db(1-c)ba$, the plants population dynamics is represented by the following difference equation:

$$p_{n+2} = \alpha p_{n+1} + \beta p_n \tag{4.23}$$

Let's consider this model for specific values of the parameters a, b, c, d, and the initial conditions p_1 and p_2. We are interested in the long-term behavior of the population of the plant (Figure 4.14).

Case I.

Let $a=3$, $b=0.6$, $c=0.4$, $d=0.2$, and $p_1=50$. We have

$$\alpha = cba = 0.4 \times 0.6 \times 3 = 0.72.$$

$$\beta = db(1-c)ba = 0.2 \times 0.6 \times (1-0.4) \times 0.6 \times 3 = 0.1296.$$

$$p_2 = cbap_1 = 0.4 \times 0.6 \times 3 \times 50 = 36.$$

Therefore, the model is represented by the second-order difference equation with the following initial conditions: $p_{n+2} = 0.72p_{n+1} + 0.1296p_n$, $p_1=50$, and $p_2 = 36$. We will use the function **PlantPopulation**, which is almost the same as the function **SecondOrderDE**

```
function [T, P]=PlantPopulation(a, b, c, d, p1, p2, k)
% This function iterates the second-order difference equation
% p(n)=alpha*p(n-1) + beta*p(n-2), n=3, 4, ..., k
% where alpha=cba and beta=db(1-c)ba
% The function parameters (input) are:
%   The constants a, b, c, d; initial conditions p1 and p2; time periods k
% Function output:
%   T=sequence of time n
%   P=sequence of p(n)

T=1:k;
P=zeros(1,k);
P(1)=p1; P(2)=p2;
alpha=c*b*a;
beta=d*b*(1-c)*b*a;
for j=3:k
  P(j)=alpha*P(j-1)+beta*P(j-2);
end;
plot(T, p, 'k.', T, P, 'k');
xlabel('Time periods n');
ylabel('p(n) values');
return;
```

We invoke this function by entering the following code:

```
>>[T, P]=PlantPopulation (3, 0.6, 0.4, 0.2, 50, 36, 20)
```

The chart of (n, p_n), $n = 1, 2, 3, \ldots, 20$ follows, and the graph of p_n vs. n is shown in Figure 4.15. Note that p_n decreases as n increases. Eventually $p_n \to 0$.

n	1	2	3	4	5	6	7	8	9	10	...	16	17	18	19	20
p_n	50	36	32.4	28	24.4	21.2	18.4	16.0	13.9	12.1	...	5.2	4.5	3.9	3.4	3.0

This conclusion can be verified by converting the second-order difference equation into a system of two first-order linear difference equations and then determining the eigenvalues of the system and applying the results discussed in Section 3.6. Assuming that $q_n = p_{n+1}$, the difference equation $p_{n+2} = 0.72p_{n+1} + 0.1296p_n$ can be converted into the system of two equations,

$$\begin{aligned} p_{n+1} &= \qquad\qquad + q_n \\ q_{n+1} &= 0.1296p_n + 0.72q_n, \end{aligned}$$

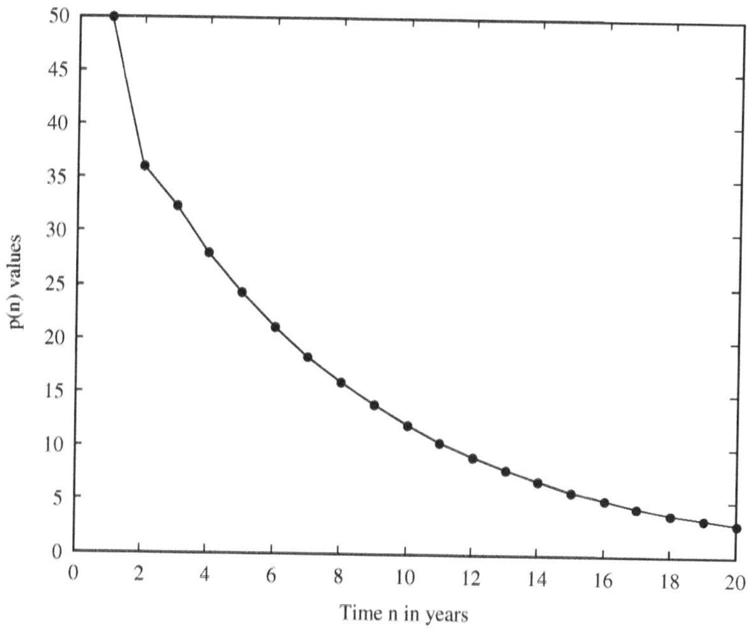

FIGURE 4.15. Graph of the number of plants in year n, p_n, vs. the time, n, in years, $n=1, 2, \ldots, 20$, where the plants' population is modeled by the second-order difference equation $p_{n+2}=0.72p_{n+1}+0.1296p_n$, with the initial conditions $p_1=50$ and $p_2=36$.

which can be represented in the matrix form

$$X_{n+1} = TX_n,$$

where $X_n = \begin{bmatrix} p_n \\ q_n \end{bmatrix}$, $T = \begin{bmatrix} 0 & 1 \\ 0.1296 & 0.72 \end{bmatrix}$, and the initial condition is $X_1 = \begin{bmatrix} 50 \\ 36 \end{bmatrix}$.

The matrix T has two eigenvalues, $\lambda_1=0.8691$ and $\lambda_2=-0.1491$, where the corresponding eigenvectors in normalized form are $V_1 = \begin{bmatrix} -0.7548 \\ -0.6560 \end{bmatrix}$ and $V_2 = \begin{bmatrix} -0.9891 \\ -0.1475 \end{bmatrix}$, respectively. Because the dominant eigenvalue $\lambda_1=0.8691<1$, $X_n \to O = \begin{bmatrix} 0 \\ 0 \end{bmatrix}$.

Case II

We will keep the values of the parameters a, b, and p_1 the same as in Case I. We will change only the values of c and d. So the parameters in this case are $a = 3$, $b = 0.6$, $c=0.6$, and $d=0.3$, with the initial condition $p_1=50$.

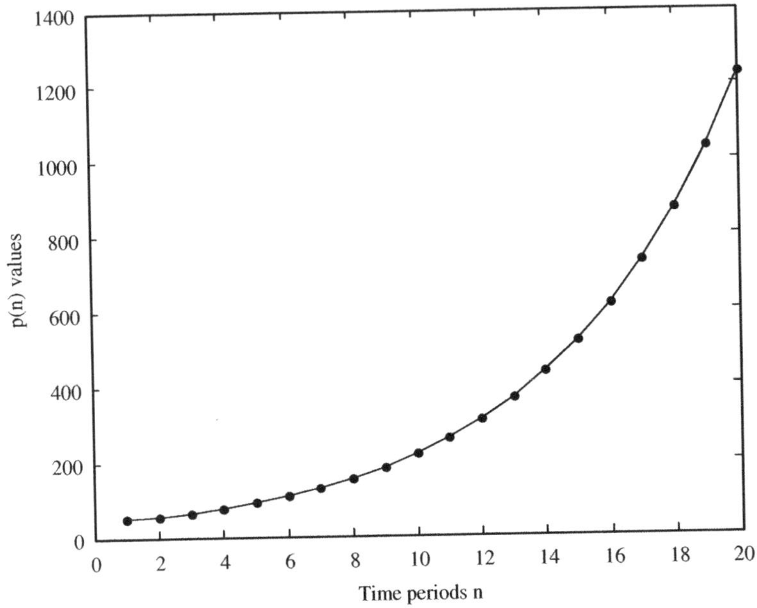

FIGURE 4.16. Graph of the number of plants in year n, p_n, vs. the time, n, in years, $n = 1, 2, \ldots, 20$, where the plants' population is modeled by the second-order difference equation $p_{n+2} = 1.08p_{n+1} + 0.1296p_n$ with the initial conditions $p_1 = 50$ and $p_2 = 54$.

Similar to Case I, we have

$$\alpha = cba = 0.6 \times 0.6 \times 3 = 1.08.$$

$$\beta = db(1-c)ba = 0.3 \times 0.6 \times (1-0.6) \times 0.6 \times 3 = 0.1296.$$

$$p_2 = cbap_1 = 0.6 \times 0.6 \times 3 \times 50 = 54.$$

The system is modeled by the second-order difference equation

$$p_{n+2} = 1.08p_{n+1} + 0.1296p_n.$$

$$p_1 = 50 \quad \text{and} \quad p_2 = 54.$$

The first 20 ordered pairs (n, p_n) are shown in the following chart and are graphed in Figure 4.16.

n	1	2	3	4	5	6	7	8	...	17	18	19	20
p_n	50	54	64.8	77	91.5	108.8	129.4	153.9	...	730.8	868.9	1,033.1	1,228.4

It can be concluded from the table and graph that p_n increases exponentially as n increases. That is, $\lim p_n \to \infty$.

Similar to Case I, we will use the eigenvalues methods to study the long-term behavior of the plant population. The second-order difference equation can be converted into the following matrix difference equation

$$X_{n+1} = AX_n,$$

where $X_n = \begin{bmatrix} p_n \\ q_n \end{bmatrix}$ and $A = \begin{bmatrix} 0 & 1 \\ 0.1296 & 1.08 \end{bmatrix}$. The eigenvalues and the corresponding eigenvectors of A are $\lambda_1 = 1.189$, $V_1 = \begin{bmatrix} -0.6437 \\ -0.7653 \end{bmatrix}$ and $\lambda_2 = -0.109$, $V_2 = \begin{bmatrix} -0.9941 \\ 0.1084 \end{bmatrix}$. Because the dominant eigenvalue $\lambda_1 = 1.89 > 1$, $X_n \to \infty$.

4.3.5. Exercises 4.3

In Exercises 1–4, for each of the second-order difference equations with initial conditions

A. Calculate the first four terms after the initial terms.

B. Use MATLAB to create a numerical solution and graph it. Can you predict the long-term behavior of the system? Explain.

1. $x_{n+2} = x_{n+1} + 2x_n$, $x_1 = 2$ and $x_2 = 3$.

2. Fibonacci sequence, $f_{n+2} = f_{n+1} + f_n$, $f_1 = 1$ and $f_2 = 1$.

3. $x_{n+2} = 0.4x_{n+1} + 0.2x_n$, $x_1 = 100$ and $x_2 = 98$.

4. $x_{n+2} - 4x_{n+1} + 4x_n = 0$, $x_1 = 2$ and $x_2 = 4$.

In Exercises 5–8, for each of the second-order difference equations with initial conditions,

A. Convert the difference equation into a system of two first-order difference equations and represent the new system by a matrix equation.

B. Use MATLAB to find the eigenvalues and the corresponding eigenvectors to predict the long-term behavior of the system.

5. Problem in Exercise 1.

6. Problem in Exercise 2.

7. Problem in Exercise 3.

8. Problem in Exercise 4.

In Exercises 9 and 10, consider the population of female seals of a colony of seals modeled in Model 4.1 by difference equation 4.15, $P_{n+2} = aP_{n+1} + bP_n$.

A. Use MATLAB to create a numerical solution of equation 4.15 and graph it. Can you predict the long-term behavior of the system?

B. Convert the difference equation into a system of two first-order difference equations and represent the new system by a matrix equation.

C. Use MATLAB to find the eigenvalues and the corresponding eigenvectors to predict the long-term behavior of the system.

9. $a=0.4$, $b=0.6$, $P_1=104$, and $P_2=100$.

10. $a=0.49$, $b=0.48$, $P_1=200$, and $P_2=198$.

In Exercises 11 and 12, consider the population dynamics of a plant, represented by equation 4.22 in Model 4.3, with the given parameters.

A. Use MATLAB to create and graph numerical solutions of equation 4.22 for 10 years. If the plant's population is declining, determine when half of the plants remain and if the plant's population is increasing, determine when the population doubles. Then predict the long-term behavior of the system.

B. Convert the second-order difference equation into a system of two first-order difference equations and represent the new system by a matrix equation. Use MATLAB to calculate the eigenvalues and the corresponding eigenvectors and use them to determine the long-term behavior of the system. Be sure that your conclusions from parts A and B are the same.

11. Equation 4.22 with $a=2$, $b=0.75$, $c=0.6$, $d=0.4$ and $p_1=100$.

12. Equation 4.22 with $a=2$, $b=0.6$, $c=0.46$, $d=0.23$ and $p_1=100$.

CHAPTER 5

MODELING WITH NONLINEAR SYSTEMS OF DIFFERENCE EQUATIONS

As we saw in the previous two chapters many situations involving more than one dependent quantity were represented by systems of linear first-order difference equations. Also we know that a system of linear difference equations might be represented by a single matrix difference equation, and therefore, matrix algebra can be used to find an analytical solution and a numerical solution of the matrix equation.

In this chapter, we investigate situations modeled by systems of nonlinear first-order difference equations. If a situation is modeled by a single nonlinear higher-order difference equation, this nonlinear difference equation can be converted to a system of nonlinear first-order difference equations. Because matrix algebra cannot be used for systems of nonlinear equations, we will rely on using MATLAB to iterate the systems and investigate their long-term behavior.

Section 5.1 focuses on the dynamic of interacting species. In Section 5.2 we study infectious disease modeling, particularly the SIR model. In Section 5.3 we introduce modeling with second-order nonlinear difference equations.

5.1. MODELING OF INTERACTING SPECIES

In this section we explore the dynamics of interactions between two or more species represented by systems of first-order nonlinear difference equations. The main types of interactions are predator–prey, competition, and mutualism. We will focus on the predator–prey models.

Explorations of Mathematical Models in Biology with MATLAB®, First Edition. Mazen Shahin.

5.1.1. A Predator–Prey Model

Consider two species in a forest, rabbits and foxes, where foxes eat rabbits and there is enough food for rabbits. We are interested in modeling the dynamic of the interaction between the predator foxes and the prey rabbits and investigate the long-term behavior of the two species.

Because the forest is a complex ecosystem, we need to make assumptions to simplify the predator–prey model. We employ the following assumptions:

i. There is enough food for rabbits and the population of rabbits increases by a constant rate—that is, the rabbit population increases exponentially.
ii. The population of rabbits decreases as a result of the interactions between rabbits and foxes.
iii. The rabbits are the only source of food for foxes. Therefore, in the absence of rabbits the population of foxes decreases by a constant rate and dies out—that is, the fox population decreases exponentially.
iv. The population of foxes increases as a result of the interactions between rabbits and foxes.
v. The rabbits and foxes live in a closed environment, which means that there is no interaction between these two species and other species, there is no emigration or immigration, and there is no harvest or hunting.

Now let us build this predator–prey model. Let

R_n = the population of the prey rabbits at the time period n.
F_n = the population of predator foxes at time period n.

Assumption i is translated into the following difference equation:

$$R_{n+1} = R_n + aR_n,$$

where a is the natural growth rate of rabbits in the absence of foxes and $a>0$.

A reasonable way to count the interactions between the rabbits and the foxes is the product R_nF_n. Assumption ii states that as a result of the interaction between the rabbit and foxes, the population of rabbits is going to decrease by a fraction of R_nF_n, say bR_nF_n, where $b>0$ is the death rate of rabbits as a result of the presence of foxes. Therefore, the equation can be modified to the following difference equation:

$$R_{n+1} = R_n + aR_n - bR_nF_n.$$
$$R_{n+1} = (1+a)R_n - bR_nF_n. \tag{5.1}$$

Note that b is very small compared to a.

Assumption iii states that in the absence of rabbits the population of foxes decreases:

$$F_{n+1} = F_n - cF_n,$$

where c is the natural death rate of foxes in the absence of rabbits and $c>0$. According to assumption iv and as a result of the interactions between the two species, the population of foxes will increase by a fraction of R_nF_n, say dR_nF_n, where $d\geq 0$ is the growth factor of foxes due to the presence of rabbits. Therefore, the above equation is modified to the following difference equation:

$$F_{n+1} = F_n - cF_n + dR_nF_n.$$

$$F_{n+1} = (1-c)F_n + dR_nF_n. \tag{5.2}$$

The factor d must be considerably smaller than c.

Therefore, this predator–prey system is modeled by a system of two nonlinear first-order difference equations, 5.1 and 5.2.

The two values R_e and F_e are called the **equilibrium values** of system 5.1 and 5.2, if there are no changes in the rabbit and fox populations. That is, $R_{n+1}=R_n=R_e$ and $F_{n+1}=F_n=F_e$ simultaneously. Substituting these values in equations 5.1 and 5.2, we get

$$R_e = (1+a)R_e - bR_eF_e.$$

$$F_e = (1-c)F_e + dR_eF_e.$$

Simplifying and solving these equations we get the following equilibrium values:

$$R_e = \frac{c}{d} \tag{5.3}$$

and

$$F_e = \frac{a}{b}. \tag{5.4}$$

Note that in addition to the equilibrium values $(R_e, F_e)=(c/d, a/b)$, there is another pair of equilibrium values, $(R_e, F_e)=(0,0)$. This means that when the population of rabbits is 0 and the population of foxes is 0, there is no change in the rabbits and foxes population. We are not interested in these trivial equilibrium values.

Note that the difference equations 5.1 and 5.2 are nonlinear. Because there is no analytical solution for a nonlinear difference equation, we will mainly rely on using a MATLAB routine to iterate the systems of difference equations and produce graphs of R_n vs. n and F_n vs. n on the same coordinate system. These graphs are called **time-series graphs**.

Here is a possible code for the function **PredPrey_TS** that accepts the parameters a, b, c, d, R_0, F_0, and n. It returns the time-series graphs for n time periods of equations 5.1 and 5.2. That is, the graphs of populations of the prey, R_n, and predator, F_n, vs. n.

```
function [T, R, F]=PredPrey_TS (a, b, c, d, R0, F0, n);
% This function iterates the Predator–Prey model
%  R(n)=(1+a)R(n-1)–bR(n-1)F(n-1)
%  F(n)=(1-c)F(n-1)+dR(n-1)F(n-1)
% and produces the time series graphs (i.e. R(n) and F(n) vs. n)
% Function parameters (input):
%  a=the natural growth rate of the prey (rabbits)
%  b=the death rate of rabbits as a result of the presence of foxes
%  c=the natural death rate of predator (foxes)
%  d=the growth factor of foxes due to the presence of rabbits
%  R0=the initial population of the prey (rabbits)
%  F0=the initial population of the predator (foxes)
%  n=number of time periods
% Function output:
%  T=sequence of time
%  R=sequence of prey population
%  F=sequence of predator population

T=1:n;
R=zeros(1, n);
F=zeros(1, n);
R(1)=R0;
F(1)=F0;
for j=2:n
  R(j)=(1+a)*R(j-1)-b*R(j-1)*F(j-1);
  F(j)=(1-c)*F(j-1)+d*R(j-1)*F(j-1);
end;
plot(T, R, 'k.', T, F, 'ko');
xlabel ('Time n in months');
ylabel ('Predator/Prey Population');
legend ('Prey', 'Predator');
return;
```

Consider the predator–prey model 5.1 and 5.2 with a=0.04, b=0.0004, c=0.08, and d=0.0002.

To call the function **PredPrey_TS** to produce the time series graphs for a=0.04, b=0.0004, c=0.08, d=0.0002, R_0=500, F_0=120, and n=600 time periods (months), we write the following command:

```
>> [T, R, F]=PredPrey_TS (0.04, 0.0004, 0.08, 0.0002,
500, 120, 600);
```

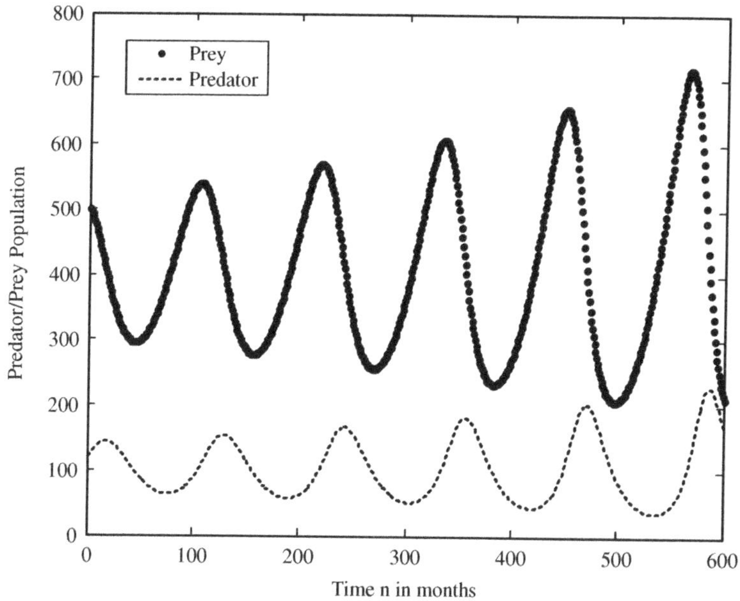

FIGURE 5.1. Time-series graphs. Graphs of the populations of prey rabbits, R_n, and the population of predator foxes, F_n, vs. the time, n, in months, $n = 0, 1, \ldots, 600$, with the initial conditions $R_0 = 500$ and $F_0 = 120$. The populations R_n and F_n are modeled by the system $R_{n+1} = R_n + 0.04R_n - 0.0004R_nF_n$ and $F_{n+1} = F_n - 0.08F_n + 0.0002R_nF_n$.

The time-series graphs are shown in Figure 5.1. From this time-series graph we see the curves for rabbits and foxes oscillate with the same period. That is, the distance (time) between two consecutive maxima for rabbits is almost the same as for the foxes. Although the two graphs oscillate with the same period, there is a shift, which means that a maximum value of the number of rabbits does not occur at the same time for a maximum value of the number of foxes. The amplitude of a wave is the vertical distance between the highest and lowest point of the wave. Figure 5.1 shows that the amplitudes of the oscillations for both rabbits and foxes increase.

Because the populations of rabbits and foxes depend on one another, it might be more informative to graph the population of foxes, F_n, vs. the population of rabbits, R_n (or the rabbits R_n vs. the foxes F_n). In this **phase-plane graph**, each point represents the number of rabbits and the number of foxes at a specific time period—that is, each point in this graph is (R_n, F_n), where the x coordinate is R_n and the y coordinate is F_n. Note that there is no explicit axis for the time. All these points will be connected to form an **orbit** (or **trajectory**) of the system.

The following is a MATLAB function, **`PredPrey_PP`**, that creates the phase plane graph for a predator–prey system, which is almost the same as PredPrey_TS.

```
function [T, R, F] = PredPrey_PP (a, b, c, d, R0, F0, n);
% Function parameters (input):
%   a = the natural growth rate of the prey (rabbits)
%   b = the death rate of rabbits as a result of the presence of foxes
%   c = the natural death rate of predator (foxes)
%   d = the growth factor of foxes due to the presence of rabbits
%   R0 = the initial population of the prey (rabbits)
%   F0 = the initial population of the predator (foxes)
%   n = number of time periods
% Function output:
%   T = sequence of time
%   R = sequence of prey population
%   F = sequence of predator population

T = 1:n;
R = zeros(1,n);
F = zeros(1,n);
R(1) = R0;
F(1) = F0;
for j = 2:n
  R(j) = (1+a)*R(j-1)  - b*R(j-1)*F(j-1);
  F(j) = (1-c)*F(j-1) + d*R(j-1)*F(j-1);
end;
plot(R, F, 'k.', R, F, 'k');
xlabel ('Prey Population');
ylabel ('Predator Population');
return;
```

Figure 5.2 is the phase-plane for this system with the same parameters and initial values of rabbits and foxes: $a = 0.04$, $b = 0.0004$, $c = 0.08$, $d = 0.0002$, $R_0 = 500$, $F_0 = 120$, and $n = 600$ time periods (months).

It is easier to interpret this phase-plane graph for understanding the dynamic of the system. The orbit is spiral, so the number of rabbits and the number of foxes goes up and down but not at the same time. We will return to the interpretation of the phase-plane after finding the equilibrium values of the system.

Using equations 5.3 and 5.4 to calculate the equilibrium values, R_e and F_e, of predator–prey system 5.1 and 5.2 with $a = 0.04$, $b = 0.0004$, $c = 0.08$, and $d = 0.0002$, we get

$$R_e = \frac{c}{d} = \frac{0.08}{0.0002} = 400.$$

$$F_e = \frac{a}{b} = \frac{0.04}{0.0004} = 100.$$

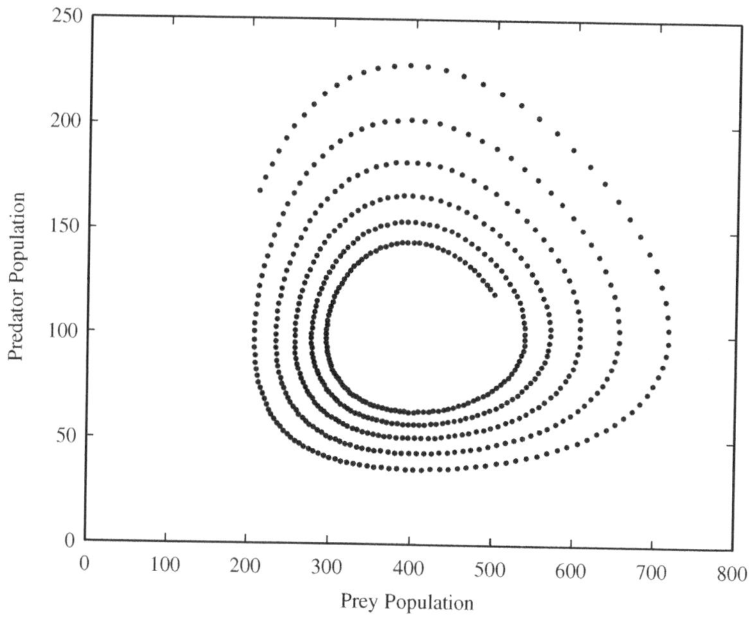

FIGURE 5.2. Phase-plane graph. Graph of the population of predator foxes, F_n, vs. the population of prey rabbits, R_n, $n = 0, 1, \ldots, 600$ months, with the initial conditions $R_0 = 500$ and $F_0 = 120$. The populations R_n and F_n are modeled by the system of difference equations $R_{n+1} = R_n + 0.04R_n - 0.0004R_nF_n$ and $F_{n+1} = F_n - 0.08F_n + 0.0002R_nF_n$.

This system has two pairs of equilibrium values, (0, 0) and (400, 100). As we mentioned before we are not interested in the equilibrium values $(R_e, F_e) = (0, 0)$, which means that there are no rabbits or foxes. If we use the initial values $R_0 = 400$ and $F_0 = 100$ (or $R_n = 400$ and $F_n = 100$), we will see there are no changes in the populations of rabbits and foxes. The time-series graphs for the system with initial values $R_0 = 400$ and $F_0 = 100$ are shown in Figure 5.3, and the phase-plane graph, which is one point (400, 100), is shown in Figure 5.4.

To determine whether the equilibrium point is stable or unstable, we find the time-series graphs (or the phase-plane) of the system with initial values very close to the coordinates of the equilibrium point. Figure 5.5 shows the time series graph for $R_0 = 401$ and $F_0 = 101$, over a long period (2000 months); note the graphs oscillate and the amplitude increases. Therefore, the equilibrium values are unstable. Figure 5.6 shows the phase plane for the system with the same parameters and initial values, $R_0 = 401$ and $F_0 = 101$. The orbit is an outward spiral, which means that the equilibrium values are unstable.

The instability of the equilibrium values may be concluded from the time-series graphs and/or phase-plane graphs of the system with the same parameters and other initial values very close to the equilibrium values, such as $R_0 = 400$ and $F_0 = 99$, $R_0 = 400$ and $F_0 = 101$, and $R_0 = 399$ and $F_0 = 99$.

Note that the horizontal line, $y = F_e = 100$, and the vertical line, $x = R_e = 400$, intersect at the equilibrium values $(R_e, F_e) = (400, 100)$ and subdivide the phase-plane into

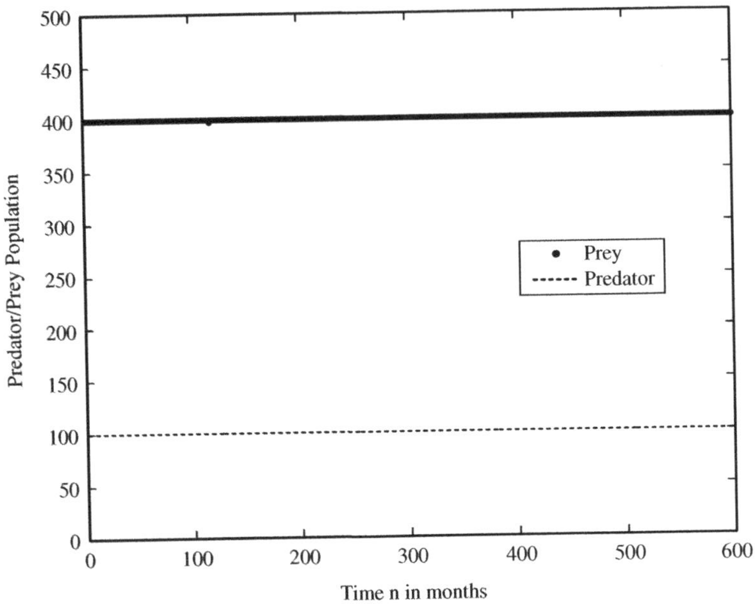

FIGURE 5.3. Time-series graphs of the system $R_{n+1}=R_n+0.04R_n-0.0004R_nF_n$ and $F_{n+1}=F_n-0.08F_n+0.0002R_nF_n$, with the initial values $R_0=400$ and $F_0=100$, for $n=0, 1, \ldots, 600$.

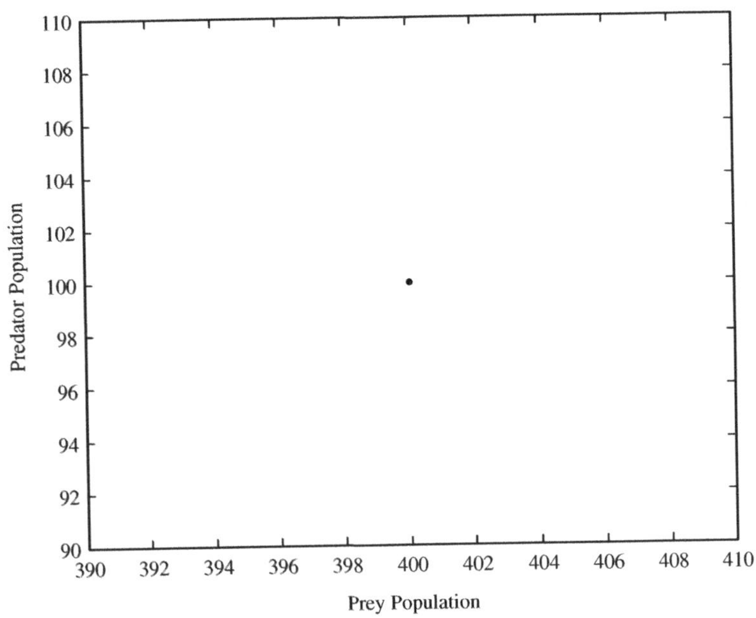

FIGURE 5.4. Phase-plane graph of the system $R_{n+1}=R_n+0.04R_n-0.0004R_nF_n$ and $F_{n+1}=F_n-0.08F_n+0.0002R_nF_n$, with the initial values $R_0=400$ and $F_0=100$, for $n=0, 1, \ldots, 600$.

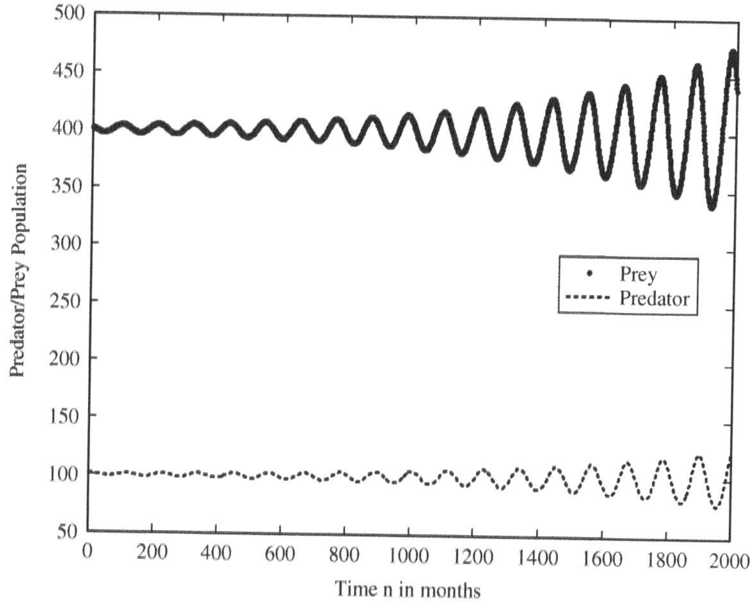

FIGURE 5.5. Time-series graphs of the system $R_{n+1}=R_n+0.04R_n-0.0004R_nF_n$ and $F_{n+1}=F_n-0.08F_n+0.0002R_nF_n$, with the initial values $R_0=401$ and $F_0=101$, for $n=0, 1, \ldots, 600$.

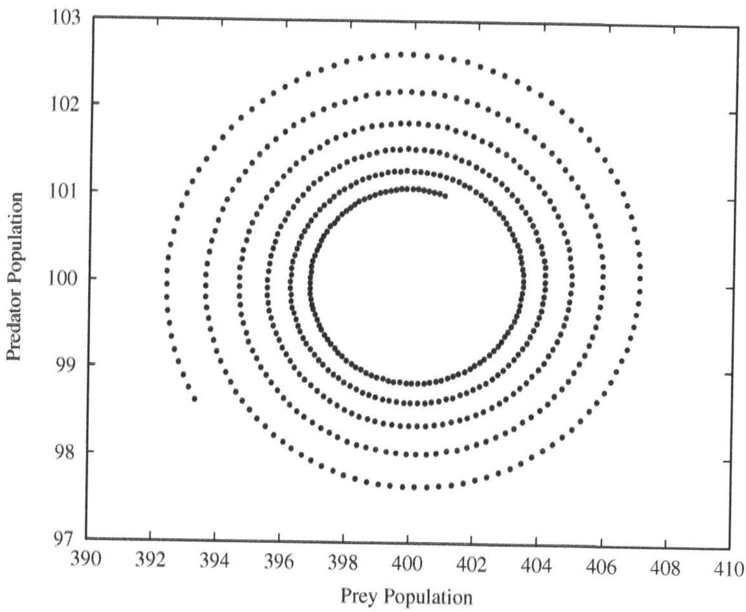

FIGURE 5.6. Phase-plane graph of the system $R_{n+1}=R_n+0.04R_n-0.0004R_nF_n$ and $F_{n+1}=F_n-0.08F_n+0.0002R_nF_n$, with the initial values $R_0=401$ and $F_0=101$, for $n=0, 1, \ldots, 600$.

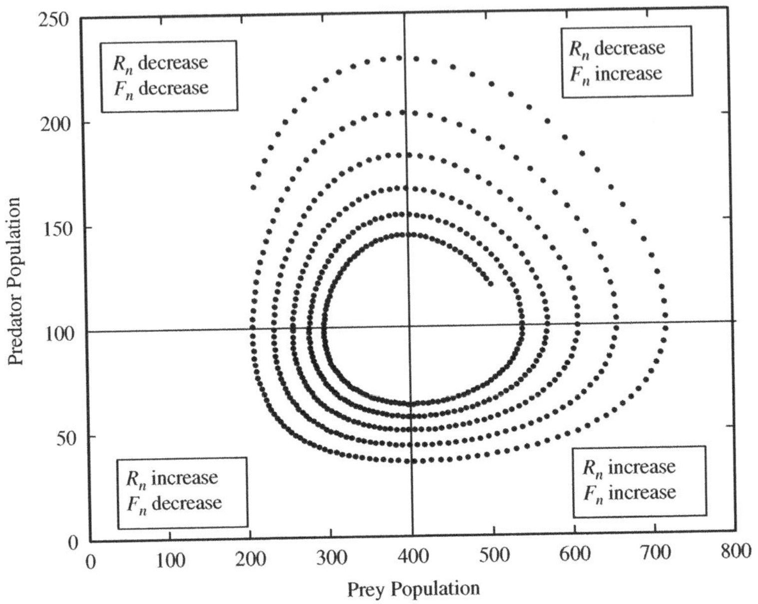

FIGURE 5.7. Figure 5.2 with the horizontal line $y=F_e=100$ and the vertical line $x=R_e=400$, where R_e and F_e are the equilibrium values of the rabbits and foxes, respectively. These two lines intersect at the equilibrium point (400, 100) and subdivide the phase plane into four quadrants. The figure shows the change of the values of R_n and F_n in each quadrant.

four quadrants. In the lower right quadrant, R_n increase and F_n increase; in the upper right quadrant, R_n decrease and F_n increase; in the upper left quadrant, R_n decrease and F_n decrease; and in the lower left quadrant, R_n increase and F_n decrease, as shown in Figure 5.7.

5.1.2. Predator–Prey Interaction with Refuges for the Prey

In addition to the assumptions of the predator–prey model represented by difference equations 5.1 and 5.2, we assume that a certain fixed number of the prey is able to hide from the predator in places called **refuges**, where the predator cannot enter and eat the prey.

Let x_n and y_n be the numbers of the prey and predator at the time period n, respectively. Let r be the number of prey who hide from the predator in refuges. If $r < x_n$, then there is an interaction between $(x_n - r)$ prey and y_n predator, whereas if $r \geq x_n$, all the prey are hidden in the refuges and consequently there is no interaction between the prey and the predator. Thus if $r<x_n$, the model is represented by

$$x_{n+1} = (1+a)x_n - b(x_n - r)y_n. \tag{5.5}$$

$$y_{n+1} = (1-c)y_n + d(x_n - r)y_n. \tag{5.6}$$

If $r \geq x_n$, the model is represented by

$$x_{n+1} = (1+a)x_n. \tag{5.7}$$

$$y_{n+1} = (1-c)y_n, \tag{5.8}$$

where a, b, c, and d are the same parameters as in the model 5.1 and 5.2. To simplify the computation, these pairs of equations may be combined in one pair of difference equations for all values of r:

$$x_{n+1} = (1+a)x_n - 0.5b\left[(x_n - r) + |x_n - r|\right]y_n. \tag{5.9}$$

$$y_{n+1} = (1-c)y_n + 0.5d\left[(x_n - r) + |x_n - r|\right]y_n. \tag{5.10}$$

5.1.3. Predator–Prey Interaction with Logistic Growth for the Prey

In the predator–prey model discussed at the beginning of this section, assumption i stated that there is enough food for rabbits and the population of rabbits increases by a constant rate—that is, the increase is exponential. This assumption is unrealistic because the limitation of food does not allow unlimited growth of rabbits. We will replace assumption i with a more realistic one.

Consider a predator–prey model with the following assumptions:

i. Due to the limitation of food for prey, the prey population follows a logistic growth model instead of an unrealistic exponential growth model.
ii. The prey's population decreases as a result of the interaction between the prey and the predator.
iii. The prey is the only source of food for predator. Therefore, in the absence of the prey, the population of the predator decreases by a constant rate and dies out.
iv. The predator's population increases as a result of the interaction between the prey and the predator.
v. The prey and the predator live in a closed environment.

Let x_n and y_n be the populations of the prey and the predator at the nth time period, respectively. Incorporating the above assumptions, this system can be modeled by the following difference equations:

$$x_{n+1} = x_n + ax_n - bx_n^2 - \alpha x_n y_n.$$

$$y_{n+1} = y_n - cy_n + \beta x_n y_n,$$

where a and b are the positive constants defined in logistic equation 2.25; α is the death rate of the prey as a result of the presence of the predator, c is the natural death rate of the predator in the absence of the prey, and β is the increase rate of the predator as a result of the interaction with the prey. Recall that the limiting value (carrying capacity) for the prey, L, is defined by $L = a/b$. Therefore, a predator–prey interaction with logistic growth of the prey may be modeled by the following nonlinear difference equations:

$$x_{n+1} = (1+a)x_n - bx_n^2 - \alpha x_n y_n. \tag{5.11}$$

$$y_{n+1} = (1-c)y_n + \beta x_n y_n. \tag{5.12}$$

5.1.4. A Competition Model

Let's consider two species, such as spotted owls and hawks, that compete for food. Consequently, the presence of one species decreases the population of the other species and the population of each species increases in the absence of the other species. Let x_n be the population of the first species (owls) at the end of n time periods and y_n be the population of the second species (hawks) at the end of n time periods. This situation can be modeled by a system of two nonlinear difference equations:

$$x_{n+1} = x_n + ax_n - bx_n y_n.$$

$$x_{n+1} = (1+a)x_n - bx_n y_n. \tag{5.13}$$

$$y_{n+1} = y_n + cy_n - dx_n y_n.$$

$$y_{n+1} = (1+c)y_n - dx_n y_n, \tag{5.14}$$

where a, b, c, and d are nonnegative constants.

The techniques developed for the predator–prey interaction and similar MATLAB functions and routines can be used to investigate competition models.

5.1.5. Mutualism Models

In a predator–prey model the prey's population decreases and the predator's population increases as a result of the prey and the predator interaction. We saw in a competition model that the populations of the two competing species decrease as a result of their interactions.

There is another type of interaction between two species called mutualism. In this interaction, each species (or organism) benefits from the other; in other words, the population of each species increases as a result of their interaction.

An example of mutualism is the association between the roots of a plant and fungus, in which the plant provides the fungus with carbohydrates and the fungus provides the plant with phosphate and nitrogen.

Assuming that the growth of each species (or organism) is exponential, the mutualism interaction of two species (or organisms) may be modeled by the following nonlinear difference equations:

$$x_{n+1} = x_n + ax_n + \alpha x_n y_n. \quad (5.15)$$

$$y_{n+1} = y_n + cy_n + \beta x_n y_n, \quad (5.16)$$

where x_n and y_n are the populations of the first and second species, respectively; a and c are the natural growth rates of first and second species, respectively; and α and β are the mutualism factor for the two species. Note that a, α, c, and β must be positive constants.

Because the growth of each species in the absence of the other species is exponential plus additional growth occurs as a result of mutualism interaction, the populations of the two species increase without bound, which is unrealistic. Because of the limitation of resources such as food and space, it would be more realistic to assume logistic growth for both species. Consequently, the mutualism model in this case may be modeled by the following nonlinear difference equations:

$$x_{n+1} = x_n + ax_n - bx_n^2 + \alpha x_n y_n. \quad (5.17)$$

$$y_{n+1} = y_n + cy_n - dy_n^2 + \beta x_n y_n, \quad (5.18)$$

where b and d are the restricted growth factors for the first and second species, respectively. The other constants are the same as for equations 5.15 and 5.16.

5.1.6. Exercises 5.1

In Exercises 1 and 2 consider the predator–prey model represented by difference equations 5.1 and 5.2 with the given parameters, where R_n and F_n are the prey and predator populations at time period n, respectively. Use the functions **PredPrey_TS** and **PredPrey_PP** to investigate the models. Do the following:

A. Find the equilibrium values of this system. Determine which equilibrium values have biological meaning.

B. Analyze the time-series graphs and phase plane graphs to predict the long-term behavior. Determine whether the equilibrium values are stable.

C. Determine the signs of the change in the prey's population, $R_{n+1} - R_n$, and the predator's population, $F_{n+1} - F_n$, in each quadrant of the axes $R = R_e$ and $F = F_e$.

D. From the time-series graphs, determine a period (i) when both populations increase, (ii) when both populations decrease, (iii) when the predator population increases and the prey population decreases, and (iv) when the prey population increases and the predator population decreases.

E. Determine the sensitivity of the system to the constants a, b, c, and d.

1. A predator–prey model between prey rabbits and predator falcons with the following parameters: $a=0.1$, $b=0.0004$, $c=0.06$, and $d=0.0001$.

2. A predator–prey model between prey rabbits and predator wolves with the following parameters: $a=0.04$, $b=0.00008$, $c=0.08$, and $d=0.00005$.

In Exercises 3 and 4 consider the predator–prey interaction with refuges for the prey modeled by difference equations 5.9 and 5.10.

A. If the equilibrium values exist, find them and determine whether they are stable.
B. Create time series and phase plane graphs to predict the long-term behavior of the system.
C. Fix the parameters a, b, c, and d, and test the system for different values of r. What conclusion can you make?

3. A predator–prey interaction between prey rabbits and predator falcons with refuges for the rabbits with the following parameters: $a=0.1$, $b=0.0004$, $c=0.06$, $d=0.0001$, and $r=200$.

4. A predator–prey interaction between prey rabbits and the predator wolves with refuges for the rabbits with the following parameters: $a=0.04$, $b=0.00008$, $c=0.08$, $d=0.00005$, and $r=300$.

In Exercises 5 and 6 consider the predator–prey interaction with logistic growth for the prey represented by equations 5.11 and 5.12 with the given parameters. Modify the functions **PredPrey_TS** and **PredPrey_PP** to create functions **PredPrey_Logistic_TS** and **PredPrey_Logistic_PP** to produce time-series and phase-plane graphs of predator–prey interactions with logistic growth for the prey models.

A. Create the time-series and phase-plane graphs to predict the long-term behavior of the system.
B. If equilibrium values exist, find them and determine whether they are stable.
C. Determine the signs of change in the prey population, $x_{n+1}-x_n$, and predator population, $y_{n+1}-y_n$, in each quadrant of the axes $x=x_e$ and $y=y_e$.

5. A system with the parameters $a=0.04$, $b=0.00002$, $\alpha=0.0004$, $c=0.08$, and $\beta=0.0002$. Select the initial values. You may start with $x_0=500$, and $y_0=120$. Iterate for $n=1000$ and $n=2000$.

6. A system with the parameters $a=0.1$, $b=0.0001$, $\alpha=0.0004$, $c=0.06$, and $\beta=0.0001$. Select the initial values. You may start with $x_0=400$, and $y_0=200$. Iterate for different values of n such as $n=300$.

In Exercises 7 and 8 consider the competition model represented by difference equations 5.13 and 5.14 with the given parameters, where x_n and y_n are the population of the first and second species at time period n, respectively. Do the following:

A. Modify the functions **PredPrey_TS** and **PredPrey_PP** to create functions **Comp_TS** and **Comp_PP** to produce time series and the phase plane graphs of competition models.

B. Find the equilibrium values x_e and y_e of the system. Determine which equilibrium values have biological meaning.

C. Use the functions **Comp_TS** and **Comp_PP** to analyze the time-series and phase-plane graphs and to predict the long-term behavior of the system. Determine whether the equilibrium values are stable or unstable.

D. Determine the signs of the change in the first species population, $x_{n+1}-x_n$, and the other species population, $y_{n+1}-y_n$, in each quadrant of the axes $x=x_e$ and $y=y_e$.

E. From the time-series graphs, determine a period (i) when both populations increase. (ii) when both populations decrease, (iii) when x_n increases and y_n decreases, (iv) when y_n increases and x_n decreases.

F. Determine the sensitivity of the system to the constants a, b, c, and d and to the initial values.

7. A competition model between owls, x_n, and hawks, y_n, that compete for the food, rats, with the parameters: $a=0.1$, $b=0.00025$, $c=0.15$, and $d=0.0003$.

8. A competition model between wolves, x_n, and foxes, y_n, that compete for the same food, rabbits, with the parameters: $a=0.4$, $b=0.0004$, $c=0.6$, and $d=0.0005$.

9. Consider a mutualism model represented by difference equations 5.15 and 5.16. Write a MATLAB function similar to **PredPrey_TS**, call it **Mutualism_TS**, that accepts a, α, c, β, x_0, y_0, and n (the number of iterations). The function creates the time-series graphs of the mutualism model. Similarly write a function to produce a phase-plane graph. Choose reasonable values for the parameters a, α, c, and β as well as the initial values x_0 and y_0.

A. If equilibrium values exist, find them.

B. From the time-series and phase-plane graphs describe the general behavior and the long-term behavior of the system.

C. Determine whether the equilibrium values are stable.

10. Repeat Exercise 9 for a mutualism model represented by difference equations 5.17 and 5.18.

5.2. THE SIR MODEL OF INFECTIOUS DISEASE

Let's consider a population in which a contagious disease is spreading. We assume that the modeling time scale is short, and for simplicity we assume the population size, N, is constant. That is, we ignore births, disease-unrelated deaths, immigration, and emigration. The population consists of three groups:

Susceptible group: those who may catch the disease are not currently infected.

Infected group: individuals who are currently infected with the disease and are contagious.

Removed group: individuals who are no longer infectious. They are either recovered, acquired immunity, or have died.

The SIR model is a simple model of the spread of contagious disease that uses the three classes *susceptible*, *infected* and *removed* (or *recovered*). The individuals of the population in the SIR model progress through the three classes in order, as illustrated in Figure 5.8.

A susceptible individual remains a susceptible or becomes infected; an infected individual stays infected for a period of time until recovered or died and removed; and a removed individual is not going to be a susceptible or infected individual.

We will use the following notations:

S_k = susceptible population at the end of time period k.
I_k = infected population at the end of time period k.
R_k = removed population at the end of time period k.

Our assumption that the population under consideration is constant can be translated into

$$S_k + I_k + R_k = N \quad \text{for all nonnegative integers } k \tag{5.19}$$

We need to develop equations to calculate S_k, I_k, and R_k.

A susceptible individual becomes infected when that individual contacts an infected individual and the disease is transmitted to the susceptible individual. The number of infected individuals depends on the frequency of the interactions between susceptible and infected individuals. It is reasonable to have the product $S_k I_k$ represent the possible number of encounters between susceptible and infected individuals.

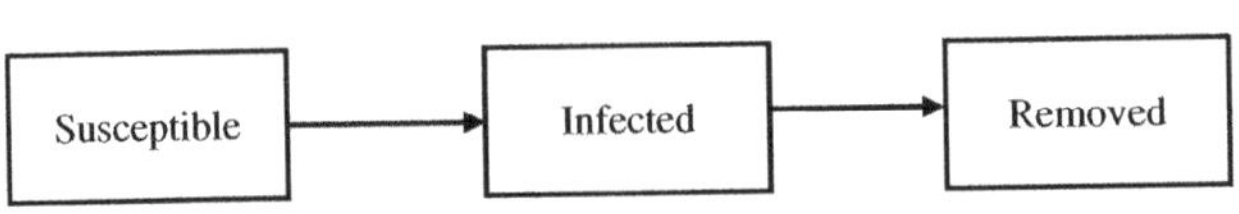

FIGURE 5.8. Progress of individuals of the population in a SIR model.

Note that we assume there is a homogenous interaction among the susceptible individuals and infected individuals. However, not every interaction between an infected individual and a susceptible individual will result in infecting the susceptible individual. Let's use α to represent the probability that the interaction between a susceptible individual and infected individual will result in a new infection. Consequently, the number of susceptible individuals who become infected as a result of interaction with infected individuals is $\alpha S_k I_k$. This number is subtracted from the number of susceptible individuals and added to the number of infected individuals; α is called the **infection rate** or the **transmission coefficient**. Therefore the dynamic of susceptible individuals can be modeled by the following difference equation:

$$S_{k+1} = S_k - \alpha S_k I_k. \tag{5.20}$$

During the time period k, the number of infected individuals will increase by $\alpha S_k I_k$ and will decrease by the infected individuals who have recovered or die. We will use β to represent the fraction of infected individuals who recover or die and consequently will be removed from the infected class; β is called **recovery rate** or **removal rate**. Therefore, the dynamic of infected class can be represented by the following difference equation:

$$I_{k+1} = I_k + \alpha S_k I_k - \beta I_k. \tag{5.21}$$

Because the number of infected individuals is decreased by βI_k, the number of removed individuals is increased by the same number. Consequently, the dynamic of the removed class is represented by the following difference equation:

$$R_{k+1} = R_k + \beta I_k. \tag{5.22}$$

The SIR model is represented by the following system of difference equations 5.19 to 5.22:

$$\begin{aligned}
S_k + I_k + R_k &= N. \\
S_{k+1} &= S_k - \alpha S_k I_k. \\
I_{k+1} &= I_k + \alpha S_k I_k - \beta I_k. \\
R_{k+1} &= R_k + \beta I_k.
\end{aligned}$$

5.2.1. Exploration 5.1

Consider the SIR model equations 5.19 to 5.22 for the spread of influenza among the students of a small college dorm with a population of 100 with the parameters: $\alpha = 0.005$ and $\beta = 0.1$. Assume that 1 student returns to the college at the beginning of the spring semester infected with influenza. Calculate S_k, I_k, and R_k for $k = 1, 2, \ldots, 50$ days; graph S_k, I_k, and R_k vs. k on one coordinate system; and describe the three graphs.

Discussion

The following is a MATLAB function called SIR to calculate S_n, I_n, and R_n and graph them vs. n. The function accepts the parameters **i_rate, r_rate, I0, N,** and **t** , where

i_rate = the infection rate (transmission coefficient α).
r_rate = removal (recovery or death) rate β.
I0 = the initial number of infected persons I_0.
N = total population.
t = number of iterations.

```
function [T, S, I, R] = SIR(i_rate, r_rate, I0, N, t)
% Function input (parameters):
%   i_rate = infection rate
%   r_rate = recovery rate
%   I0 = initial number of infected individuals
%   N = total number of population
%   t = time
% Function output:
%   T = sequence of time
%   S = sequence of susceptible individuals
%   I = sequence of infected individuals
%   R = sequence of recovered individuals

T = 1:t;
S = zeros(1, t);
I = zeros(1, t);
R = zeros(1, t);
S(1) = N - I0;
I(1) = I0;
R(1) = 0;
for k = 2:t
  S(k) = S(k-1) - i_rate * S(k-1) * I(k-1);
  I(k) = I(k-1) + i_rate * S(k-1) * I(k-1) - r_rate * I(k-1);
  R(k) = R(k-1) + r_rate * I(k-1);
end;
plot(T, S, 'k.', T, I, 'ko', T, R, 'k*');
hold on;
plot(T, S, 'k', T, I, 'k', T, R, 'k');
hold off;
xlabel('Time n in days');
```

```
ylabel('S(n), I(n), and R(n)');
legend('Susceptible', 'Infected', 'Recovered');
```

To call the function SIR with the parameters i_rate = 0.005, r_rate=0.1, I0=1, N = 100, and t=50 days, write the code

```
>> [T, S, I, R]=SIR (0.005, 0.1, 1, 100, 50);
```

The graphs of S_n, I_n, and R_n vs. n with $n = 0, 1, 2, \ldots, 50$ are shown in Figure 5.9.

The graphs show that the following:

The number of infected persons increases and then decreases. The peak value of infected persons is about 51, and it occurs on day 17—that is $I_{17} = 51$. We have $I_{50} \cong 2$.

The number of susceptible persons is decreasing, and it reaches approximately 1 at day 31—that is, $S_{31} \cong 1$.

The number of recovered persons is increasing, and we have $R_{50} \cong 97$.

In the exercises you will investigate the effect of the system parameters and the initial values on the behavior of the model.

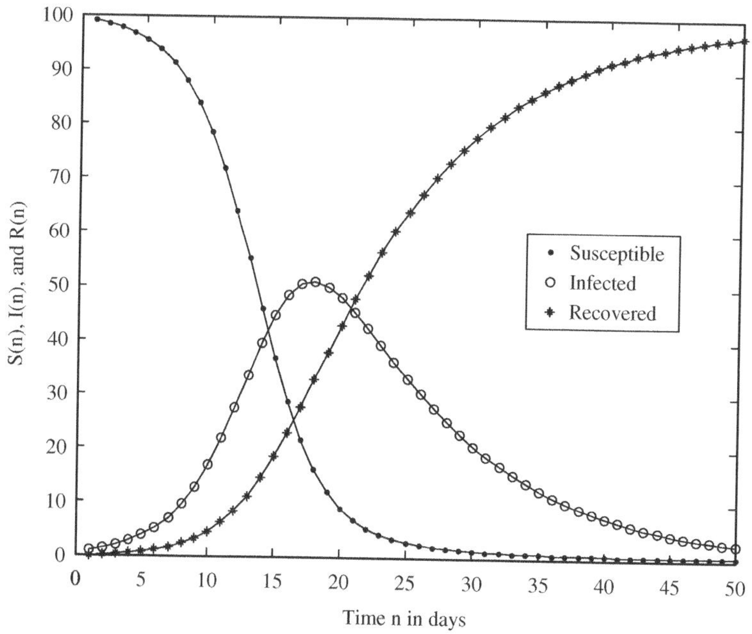

FIGURE 5.9. Time-series graphs of a SIR model. Graphs of susceptible, infected, and recovered populations S_n, I_n, and R_n respectively vs. time, n, in days. The populations are modeled by the system of difference equations: $S_{k+1} = S_k - 0.005S_kI_k$, $I_{k+1} = I_k + 0.005S_kI_k - 0.1I_k$, and $R_{k+1} = R_k + 0.1I_k$ with $I_0 = 1$, and $k = 1, 2, \ldots, 50$ days.

Transmission Coefficient and Recovery Rate

How are the transmission coefficient and recovery rate determined? This is an interesting question. Because the recovery (removal) rate β is the portion of infected persons removed each time period, the value of β depends on the duration of time a person stays infected. In general,

$$\beta = \frac{1}{\text{Average duration of infectiousness}}. \tag{5.23}$$

For example, if the average duration of a certain flu infection is 10 days, then $\beta = 1/10 = 0.1$.

Knowing the initial number of infected persons, I_0, and the number of infected persons at time period 1, the transmission coefficient, α, can be determined using equation 5.20. Equation 5.20 with $k=0$ is $S_1 = S_0 - \alpha S_0 I_0$, where $\alpha S_0 I_0$ is the number of newly infected persons in time period 1. The parameter α is determined from the following equation:

$$\text{Number of new infected persons in time period 1} = \alpha S_0 I_0$$

so

$$\alpha = \frac{\text{Number of new infected persons in time period 1}}{S_0 I_0}.$$

Because $S_0 = (N - I_0)$, we have

$$\alpha = \frac{\text{Number of new infected persons in time period 1}}{(N - I_0) I_0}. \tag{5.24}$$

For example, if the total population is 100 and the initial number of sick persons is 2 and the number of newly infected persons on day 1 is 3, we get

$$\alpha = \frac{3}{(100-2)\cdot 2} = \alpha = \frac{3}{(100-2)\cdot 2} = 0.0153.$$

Note that equation 5.24 is a simple method for determining α, but if the number of newly infected persons in time period 1 is not accurate, then α is not accurate.

In addition to the SIR model, we will discuss one similar model of the spread of contagious diseases.

5.2.2. SIS Model of Infectious Disease

In the SIR model we assumed that recovered individuals had acquired immunity and thus are not susceptible. Now we assume that for certain diseases, such as the common cold, a recovered person is not immune from getting the disease again.

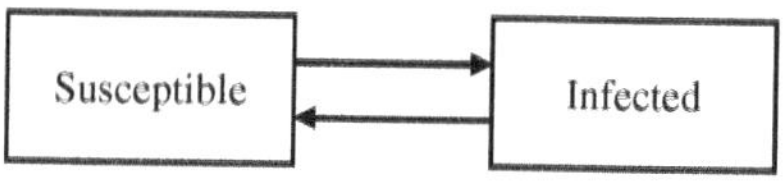

FIGURE 5.10. Progress of individuals of the population in a SIS model.

Therefore, a recovered person becomes susceptible again, and we get the so-called **SIS** model, as illustrated in Figure 5.10.

This model can be represented by equation 5.19 and the following:

$$S_{k+1} = S_k - \alpha S_k I_k + \beta I_k. \quad (5.25)$$

$$I_{k+1} = I_k + \alpha S_k I_k - \beta I_k. \quad (5.26)$$

5.2.3. Exercises 5.2

In Exercises 1–7 consider the SIR model represented by equations 5.19 to 5.22 with given parameters. For each exercise, investigate the model by graphing S_n, I_n, and R_n vs. n. Describe the model behavior and the effect of the values of the parameters. Let the total population be $N=100$.

1. Recovery rate > infection rate (e.g., $\alpha=0.01$, $\beta=0.1$, $I_0=1$).

2. Recovery rate < infection rate (e.g., $\alpha=0.01$, $\beta=0.008$, $I_0=1$).

3. Recovery rate = 0 (e.g., $\alpha=0.01$, $\beta=0$, $I_0=1$).

4. Recovery rate = infection rate (e.g., $\alpha=0.01$, $\beta=0.01$, $I_0=1$).

5. Similar to Exercise 1 but $I_0=3$—that is, recovery rate > infection rate (e.g., $\alpha=0.01$, $\beta=0.1$, $I_0=3$).

6. Similar to Exercise 2 but $I_0=3$—that is, recovery rate < infection rate (e.g., $\alpha=0.01$, $\beta=0.001$, $I_0=3$).

7. Recovery rate >> infection rate (e.g., $\alpha=0.001$, $\beta=0.1$, $I_0=1$).

In Exercises 8–14 consider the SIS model represented by equations 5.19, 5.25. and 5.26 with given parameters. For each exercise, investigate the model by graphing S_n, I_n, and R_n vs. n. Describe the model behavior and the effect of the values of the parameters. Let the total population be $N=100$.

8. Same as in Exercise 1.

9. Same as in Exercise 2.

10. Same as in Exercise 3.

11. Same as in Exercise 4.

12. Same as in Exercise 5.

13. Same as in Exercise 6.

14. Same as in Exercise 7.

15. Consider the SIR model with one modification. A constant fraction of the infected persons are quarantined to decrease the interactions between susceptible and infested persons. Consequently the severity of the epidemic is decreased. Let q be the fraction of the quarantined infected persons.

A. Modify the model. *Hint*: The infection rate, α, in SIR model should be changed to $(1-q)\alpha$; therefore, newly infected individuals are $(1-q)\alpha S_n I_n$.

B. Test the new model for the following parameters: $N=500$, $I_0=20$, $\alpha=0.001$, $\beta=0.45$, and $q=0.2$ (i.e., 20% of infected people are quarantined).

C. Compare the SIR model without quarantine ($N=500$, $I_0=20$, $\alpha=0.001$, and $\beta=0.45$) and the modified model with quarantine tested in part B. Describe the difference between the two models.

D. Investigate and describe the effect of increasing or decreasing the initial value of infected persons, I_0. *Hint*: Fix all other parameters and change I_0.

5.3. MODELING WITH SECOND-ORDER NONLINEAR DIFFERENCE EQUATIONS

In this section we are interested in some models represented by second-order nonlinear difference equations such as delayed logistic models.

Because, in general, there is no analytical solution of a second-order nonlinear difference equation, we will rely on obtaining a numerical solution of the equation or of the equivalent system of two first-order nonlinear difference equations.

The following example illustrates the iteration process of a second-order nonlinear difference equation for obtaining a numerical solution as well as converting the equation into a system of two first-order difference equations.

Example 5.1

Consider the following nonlinear second-order difference equation:

$$x_{n+2} - 4x_{n+1} + 2x_n x_{n+1} = 0,$$

with the initial conditions $x_0=1$ and $x_1=3$.

i. Find x_2, x_3 and x_4.

ii. Convert the second-order equation into a system of first-order difference equations.

iii. Use MATLAB to iterate the second-order difference equation and the system of equations.

Solution

i. The given difference equation can be written in the following form:

$$x_{n+2} = 4x_{n+1} - 2x_n x_{n+1} \tag{5.27}$$

Iterating equation 5.27 with the initial condition $x_0 = 1$ and $x_1 = 3$, we get

$$x_2 = 4x_1 - 2x_0 x_1 = 4(3) - 2(1)(3) = 6$$
$$x_3 = 4x_2 - 2x_1 x_2 = 4(6) - 2(3)(6) = -12$$
$$x_4 = 4x_3 - 2x_2 x_3 = 4(-12) - 2(6)(-12) = 96.$$

ii. We introduce a new dependent variable, y_n,

$$y_n = x_{n+1}.$$

Note that $y_{n+1} = x_{n+2}$. Substitute in equation 5.27 for x_{n+1} and x_{n+2}, and we get

$$y_{n+1} = 4y_n - 2x_n y_n.$$

Therefore, we have the following system of two first-order difference equations:

$$x_{n+1} = y_n.$$
$$y_{n+1} = 4y_n - 2x_n y_n.$$

We need to convert the initial conditions x_0 and x_1 into initial conditions x_0 and y_0. We have $x_0 = 1$. Letting $n = 0$ in equation $y_n = x_{n+1}$, we get $y_0 = x_1 = 3$. Therefore, the new initial conditions are $x_0 = 1$ and $y_0 = 3$.Iterating the system with the initial conditions $x_0 = 1$ and $y_0 = 3$, we get

$$x_1 = y_0 = 3$$
$$y_1 = 4y_0 - 2x_0 y_0 = 4(3) - 2(1)(3) = 6$$
$$x_2 = y_1 = 6$$
$$y_2 = 4y_1 - 2x_1 y_1 = 4(6) - 2(3)(6) = -12.$$
$$x_3 = y_2 = -12$$
$$y_3 = 4y_2 - 2x_2 y_2 = 4(-12) - 2(6)(-12) = 96$$

Note that x_0, x_1, x_2, and x_3 are the same as the result of iterating the second-order difference equation.

iii. The following is a possible code to iterate the second-order difference equation:

$$x_{n+2} = 4x_{n+1} - 2x_n x_{n+1}, \quad n = 1, 4, \ldots, 20,$$

with the initial conditions $x_1 = 1$ and $x_2 = 3$. Note that we start with x_1 instead of x_0.

```
T=1:20;
X=zeros(1, 20);
X(1)=1;
X(2)=3;
for k=3:20
  X(k)=4*X(k-1)-2*X(k)*X(k-2);
end;
plot(T, X);
```

5.3.1. Model 5.1: A Nonlinear Delayed Logistic Model

We assume that for certain species there is a maturation period for sexual maturity, T. Therefore, the population model must include a delay effect. The population P_{n+1} depends on P_n and P_{n-T}.

Let's consider some population models in which the maturity period $T=1$. That is, each member of the population becomes productive in its second year and is able to produce new individuals each year thereafter. In this case, each member of the population at time $(n-1)$ and n contribute to the population at time $(n+1)$.

In Section 2.3 we discussed the logistic growth model

$$P_{n+1} = P_n + aP_n - bP_n^2, \tag{5.28}$$

where P_n is the population of the species after n time periods, a is the unrestricted growth rate, and b is the inhibiting constant. Letting $x_n = \dfrac{b}{1+a}P_n$, equation 5.28 is scaled and transformed into

$$x_{n+1} = rx_n(1-x_n), \tag{5.29}$$

where $r=(1+a)$ is a parameter and $0<x_0<1$.

In this logistic model, it was assumed that every member of the species in the nth generation contributes to the population in the $(n+1)$st generation. This assumption is true for certain species, such as most insects. However, it is not the case for many other species where there is a substantial maturation period (such as for baleen whales) or species that migrate for breeding. Consequently, the population model must incorporate a delay effect. In the **delayed logistic** population model, we assume that the $(n+1)$st population depends on both the nth and $(n-1)$st populations. One reasonable modification of equation 5.28 for incorporating the delay effect is

$$P_{n+1} = P_n + aP_n - bP_nP_{n-1}. \tag{5.30}$$

Letting $x_n = \frac{b}{1+a}P_n$, equation 5.30 is scaled and transformed into

$$x_{n+1} = rx_n(1 - x_{n-1}),$$

where $r = (1+a)$.

The last equation can be written in the following form:

$$x_{n+2} = rx_{n+1}(1 - x_n). \tag{5.31}$$

Equation 5.31 is a second-order nonlinear difference equation. Because, in general, there is no analytical solution of this nonlinear equation, we will investigate its numerical solution for $0 < x_0 \le 1$ (note that equation 5.31 is scaled).

5.3.2. Model 5.2: A Nonlinear Delayed Population

In Section 4.3 we considered a population model in which populations P_{n+1} and P_n contribute to population P_{n+2}. That situation was modeled by

$$P_{n+2} = \alpha P_{n+1} + \beta P_n,$$

where α and β are constants representing growth factors. Now let's modify this model by assuming that α and β are not constants but are functions of the sum $P_{n+1} + P_n$. It is reasonable to assume that if the sum $P_{n+1} + P_n$ equals the limiting value (or the carrying capacity) L, the contributions from P_{n+1} and P_n to P_{n+2} are zero. Therefore, α and β are replaced by the following functions:

$$a\left(1 - \frac{P_{n+1} + P_n}{L}\right).$$

$$b\left(1 - \frac{P_{n+1} + P_n}{L}\right).$$

Consequently, this situation is modeled by the following difference equation:

$$P_{n+2} = a\left(1 - \frac{P_{n+1} + P_n}{L}\right)P_{n+1} + b\left(1 - \frac{P_{n+1} + P_n}{L}\right)P_n,$$

which simplifies to

$$P_{n+2} = \left(1 - \frac{P_{n+1} + P_n}{L}\right)(aP_{n+1} + bP_n). \tag{5.32}$$

Equation 5.32 is a second-order nonlinear difference equation.

5.3.3. Model 5.3: Delayed Nonlinear Competing Species

In Section 5.1 we considered a competition model, in which two species, such as owls and hawks, are competing for the same food. The model was represented by a system of two nonlinear first-order difference equations:

$$x_{n+1} = (1+a)x_n - bx_x y_n.$$

$$y_{n+1} = (1+c)x_n - dx_x y_n.$$

Now we will modify this model with a delay effect due to the gestation period for the species. We assume that the populations x_n and x_{n-1} contribute to the population x_{n+1}; therefore, in the absence of the competitor species we have

$$x_{n+1} = ax_n + bx_{n-1},$$

where $a>0$ and $b>0$ are the fractions of the current population x_n and the previous population x_{n-1} that become members of population x_{n+1}.

In the presence of the competitor, this equation would be modified to

$$x_{n+1} = ax_n + bx_{n-1} - cx_n y_n,$$

where $c \geq 0$ represents the interaction factor between the two species that results in a decrease in the population x_{n+1}. Similarly, the second species is modeled by the following difference equation:

$$y_{n+1} = dy_n + ey_{n-1} - fx_n y_n,$$

where $d>0$, $e>0$, and $f \geq 0$ are parameters for the second species and similarly to a, b, and c, respectively.

For convenience we will write the last two equations as

$$x_{n+2} = ax_{n+1} + bx_n - cx_{n+1} y_{n+1}. \tag{5.33}$$

$$y_{n+2} = dy_{n+1} + ey_n - fx_{n+1} y_{n+1}. \tag{5.34}$$

Equations 5.33 and 5.34 are nonlinear second-order difference equations.

5.3.4. Exercises 5.3

1. Consider the delayed logistic model 5.31. For each of the following cases, use MATLAB to create the numerical solutions (n, x_n) and graph them; then predict the long-term behavior.

Case 1: Fix the initial values $x_0=0.1$ and $x_1=0.2$. Select at least four different values of r between 1.5 and 2.5.

Case 2: Select one value for r ($1.5<r<2.5$). Fix r and change the initial conditions.

2. Consider the delayed logistic growth model 5.31.

A. Convert the second-order difference equation 5.31 with the initial conditions $x_0=0.1$ and $x_1=0.2$ into a system of two first-order difference equations.

B. Use MATLAB to find the numerical solution of the converted system. Be sure that the obtained solution is the same as for Exercise 1, with the same parameter r and initial conditions. Can you predict the system's long-term behavior? Explain your answer.

3. Modify the logistic growth model 5.31 to incorporate an immigration of a constant C per period. Use MATLAB to investigate this model for different values of C.

4. Modify the logistic growth model 5.31 to incorporate a migration or hunting/harvesting of a constant C per period. Use MATLAB to investigate this model for different values of C.

5. Determine the equilibrium value of the delayed logistic model 5.31 that has biological meaning in terms of r. Note that $0<x_n\leq 1$. Determine the equilibrium value of the logistic delayed models developed in Exercise 3, and use MATLAB to test these equilibrium values as well as their stability/instability.

6. Consider the nonlinear delayed model 5.32. For each of the following cases, use MATLAB to create the numerical solutions (n, P_n) and graph them; then predict the long-term behavior.

Case 1: Fix the initial values $P_0=10$ and $P_1=12$. Select at least four different values of a, b, and L, e.g., $a=0.6$, $b=0.4$, and $L=800$.

Case 2: Select one set of a, b, and L. Fix these values and change the initial conditions. Use MATLAB to investigate these situations.

7. Consider the delayed logistic growth model 5.32.

A. Convert the second-order difference equation 5.32 with the initial conditions $x_0=10$ and $x_1=12$ into a system of two first-order difference equations.

B. Use MATLAB to find the numerical solution of the converted system. Be sure he obtained solution is the same as for Exercise 6, with the same parameters and initial conditions. Can you predict the system's long-term behavior? Explain your answer.

8. Modify the logistic growth model 5.32 to incorporate an immigration of a constant C per period. Use MATLAB to investigate this model for different values of C.

9. Modify the logistic growth model 5.32 to incorporate a migration or hunting/harvesting of a constant C per period. Use MATLAB to investigate this model for different values of C.

10. Determine the equilibrium value of the delayed logistic model 5.32 that has biological meaning in terms of a, b, and L. Determine the equilibrium value of the logistic delayed models developed in Exercises 3 and 4. Use MATLAB to test these equilibrium values.

11. Consider the delayed competition model 5.33 and 5.34 with the following parameters: $a=0.75$, $b=0.46$, $c=0.0002$, $d=0.82$, $e=0.53$, and $f=0.00015$. Find the equilibrium values for the system.

12. Use MATLAB with different initial values to produce the time series graphs and phase plane graphs for the system in Exercise 5. Determine the long-term behavior of the system. Is the system sensitive to the initial values? Explain your answer.

REFERENCES

Allman, Elizabeth S., and John A. Rhodes. *Mathematical Models in Biology: An Introduction.*: Cambridge University Press, Cambridge, 2004.

Anton, Howard and Chris Rorres. *Elementary Linear Algebra: Application Version.* 10th ed.: Anton Textbooks, Cherry Hill, NJ, 2010.

Arney, David C., Frank R. Giordano, and John S. Robertson. *Discrete Dynamical Systems: Mathematics, Methods, and Models.*: McGraw-Hill, New York, 2001.

Donnelly, Karen. *MATLAB Manual: Computer Laboratory Exercises.*: Saunder College, Philadelphia, PA, 1995.

Edelstein-Keshet, Leah. *Mathematical Models in Biology.*: Society for Industrial and Applied Mathematics, Philadelphia, PA, 2005.

Elaydi, Saber N. *An Introduction to Difference Equations.* 3rd ed.: Springer-Verlag, New York, 2010.

Fulford, Glenn, Peter Forrester, and Arthur Jones. *Modelling with Differential and Difference Equations.*: Cambridge University Press, Cambridge, 1997.

Goldberg, Samuel. *Introduction to Difference Equations*: Dover, New York, 1986.

Larson, Ron. *Elementary Linear Algebra.* 7th ed.: Brooks/Cole, Cengage Learning, Boston, MA, 2013.

Lay, David C. *Linear Algebra and Its Applications.* 4th ed.: Pearson Education, Boston, MA, 2012.

Leon, Steven J. *Linear Algebra with Applications.* 8th ed.: Pearson Prentice Hall, Upper Saddle River, NJ, 2009.

Explorations of Mathematical Models in Biology with MATLAB®, First Edition. Mazen Shahin.

Marotto, Frederick R. *Mathematical Modeling Using Discrete Dynamical Systems.*: Brooks/Cole, Belmont, CA, 2005.

Mooney Douglas D., and Randall J. Swift, *A Course in Mathematical Modeling.*: Mathematical Association of America, Washington, D.C., 1999.

Poole, David. *Linear Algebra: A Modern Introduction.* 2nd ed.: Thomson Brooks/Cole, Boston, MA, 2006.

Pratap, Rudra. *Getting Started with MATLAB: A Quick Introduction for Scientists and Engineers.*: Oxford University Press, Oxford, 2009.

Sandefur, James T. *Discrete Dynamical Modeling.*: Oxford University Press, Oxford, 1993.

Williams, Gareth. *Linear Algebra with Applications.* 8th ed.: Jones & Bartlett, Burlington, MA, 2012.

INDEX

Explorations of Mathematical Models in Biology with MATLAB®, First Edition. Mazen Shahin.

Printed and bound by CPI Group (UK) Ltd, Croydon, CR0 4YY
16/09/2026
14963413-0003